Innovative Technologies and Signal Processing in Perinatal Medicine

Danilo Pani • Chiara Rabotti
Maria Gabriella Signorini • Laura Burattini

Editors

Innovative Technologies and Signal Processing in Perinatal Medicine

Volume 1

Editors
Danilo Pani
Department of Electrical and Electronic
Engineering
University of Cagliari
Cagliari, Italy

Maria Gabriella Signorini
Department of Electronic, Information
and Bioengineering, DEIB
Politecnico di Milano
Milan, Italy

Chiara Rabotti
Family Care Solutions
Philips Research
Eindhoven, The Netherlands

Laura Burattini
Department of Information Engineering
Università Politecnica delle Marche
Ancona, Italy

ISBN 978-3-030-54405-8 ISBN 978-3-030-54403-4 (eBook)
https://doi.org/10.1007/978-3-030-54403-4

This Springer imprint is published by the registered company Springer Nature Switzerland AG
The registered company address is: Gewerbestrasse 11, 6330 Cham, Switzerland

Foreword

In an age when people scan through hyperlinked digital literature, copying and pasting references and text at a break-neck pace, shooting salvos of papers at the sea of the ever-increasing population of journals, it may seem strange to compile a scientific book (or more accurately, a *monograph*). Does anyone ever sit down and read a textbook from cover to cover, or pull a text from a shelf to look up a key fact anymore? I still do, but I may be showing my age now. Yet, I lived through the internet revolution, and although I may look fondly back to the months I sat on the basement floors of the Radcliffe Science Library in Oxford until I finally had the epiphany my thesis required, the digitization and democratization of scientific literature, together with the rapid changes in computational power, has transformed the rate at which I can survey the field, generate and test new ideas. Nevertheless, there is still a place in the world for a good book. Periodically, each field needs a pinch point for information – somewhere to tie together the wild strains of through, synthesize them into a volume to provide perspective, and challenge the researcher to push forward in the most promising directions. It also presents the field in a less fractured way, allowing the reader to trace a continuum through the field and see beyond the text and make connections to other fields.

While the world of perinatal monitoring shares much with the pediatric and adult monitoring literature, it is also a unique space. The transition from fetus to newborn is a perilous path in which the human turns from something unrecognizable into a (mostly) self-regulating being, with the ability to modify breathing and cardiac function in response to the environment, react to light, temperature, touch, speech, and many other stimuli. If all goes well, the moment of birth sees an incredible change in circulation, with the umbilical cord being clamped, preventing the delivery of oxygen and nutrients from the mother. With the first breaths of air (which previously were just for practice), the alveoli in the lungs are cleared of fluid, and the resultant blood and pulmonary pressure changes cause blood to flow to the lungs to acquire the vital oxygen the fetus needs. Seeing their tissue perfuse from blue to red is a marvelous sight to witness this moment, especially in your own children. Over the first few months of life, as most parents are acutely aware, the cardiovascular system continues to evolve, with changes in autonomic tone, and the establishment of the circadian rhythm entrained by the sunrise and sunset. This rapid growth creates a challenge, though. Knowing the age (from conception) is essential, because a young, healthy fetus can resemble an older growth-retarded fetus in many ways (from size to autonomic complexity). Yet our measures of conception date are based mainly on guesswork. The other issue is that a fetus and newborn mature at such a rapid rate, and in such a nonlinear manner, that it's hard to compare one week with another. We think of growth curves as just that – smooth curves, but in reality, they are a series of curves, plateaus, and step changes, creating a complex tracking problem that we rarely encounter in children or adults. I've often wondered why the world of perinatal monitoring has not advanced as fast as other areas, frequently sticking to antiquated technologies and cultural myths (or "guidelines") based on gut intuition rather than hard science. Perhaps it is because of these issues. But it is also because it is the most vulnerable time in our lives, and for a good reason, as a society, we shy away from invasive and perturbative monitoring. This makes the need for hard empirical research in this field all the more important – the most vulnerable in society deserve better care. The challenge is to do this in the most compassionate and gentle way that we can.

This monograph provides a building block in that context. It leads the reader through the key monitoring technologies for perinatal monitoring, from engineering foundations to the latest research. For anyone starting research in the field of perinatal monitoring, this is an excellent desk companion to be digested in part or as a whole. The text can be considered as a resource for both the research student on their initial exploration into the field and as a reference to those further along in their studies, who may feel that it is time to step back and reflect on where their research has taken them to date. It will also be instructive for those in our field working in industry. The commercial space is in need of something new – not a shiny new wireless device that can be worn on the wrong part of the body, or an expensive cage that pins the user to the bed – but rather something that marries ground-breaking research with affordable and scalable technology to reach the parts of the world lacking resources. I encourage the reader to think about these bigger issues as they pore through the research and guidance provided in this monograph, considering both the social and economic issues, the environmental footprint of disposable and energy-hungry systems, and potential cultural issues and intrinsic biases that we accidentally build into all our creations.

I wish the reader well in their research and hope you take the time to sit back with a textbook and experience how knowledge acquisition used to happen. Without taking the time to see the bigger picture, you may find yourself drilling in the wrong location. Or as Xun Zi is thought to have said:

凡人之患，蔽于一曲，而暗于大理

("In order to properly understand the big picture, everyone should avoid becoming mentally clouded and obsessed with one small element of truth.")

Gari D. Clifford

August 14, 2020

Preface

Pregnancy is a critical time for the health of mother and fetus, with significant risks for both. Early prenatal diagnosis can enable therapeutic interventions in utero or postpartum (or childbirth programming through cesarean section) in order to minimize risks for the fetus in the short and long term. Tools for pregnancy monitoring can as well be adopted to follow the mother and baby throughout gestation, up to labor and more, in case something goes wrong. In particular, the perinatal period, spanning from the third trimester of pregnancy up to 1 month after birth, is most critical for the baby. For this reason, in the last decades, biomedical engineering supported and fostered scientific research towards identification of new models, parameters, algorithms, and tools that can improve the quality of fetal monitoring, predict the outcomes, and allow physicians to intervene in an appropriate manner to ensure a healthy future for the baby. This research led to important successes in the envisioning and designing of medical devices (including software algorithms) to support the clinicians in the diagnostics and monitoring of perinatal life critical conditions.

In this context, instrument development and signal processing research interpenetrate, and specific advanced competences are required of the professionals able to operate in this field. Even though the interest is huge, and there is a significant amount of scientific papers discussing specific aspects related to technologies and signal processing in perinatal medicine, there is a lack of comprehensive books presenting such topics with a clear educational purpose and consequently a didactic approach. This book comes from the experience at the First International Summer School on Technologies and Signal Processing in Perinatal Medicine, which was held in Pula, Italy, from July 2 to 6, 2018. The School was realized thanks to the collaboration and financial contribution of the Regione Autonoma della Sardegna and of Sardegna Ricerche and attracted students from all over the world. The book's content reflects some of the most important master lectures provided in that context by eminent scientist, expert clinicians, and technologists.

As such, it is interesting in particular for students and young researchers approaching this topic for the first time and looking for a clear introduction. Nevertheless, any researcher already active in the field could find in this book a clear representation of the presented themes, an invaluable source of references, and a stimulating analysis of different techniques presented by recognized experts in the field. By finding all the presented themes in one book, the reader has a clear introduction to the topic, including the updated references representing the state of the art in the field.

This book aims to enable the reader to operate in the context of technologies for perinatal medicine by creating product innovation through scientific research. In particular, two main technologies are considered: those descending from the adoption of ultrasounds and those based on the acquisition of electro-physiological signals.

Ultrasounds in pregnancy have been widely used in clinical practice and represent today's leading technology for antenatal diagnostics and monitoring. The key aspects related to their physics and technologies for imaging in clinical settings are presented along with their different application fields throughout the pregnancy. The complexity associated to the correct interpretation of fetal images, in particular when small morphological details such as those of the fetal heart are being studied, makes this tool strongly operator-dependent. As such, the development of frameworks for remote real-time consulting is of paramount importance for the provision of correct diagnoses in a timely manner. Moreover, the identification of ultrasound-based markers can be useful for the objective assessment of the fetus against the occurrence of several life-threatening conditions, such as intrauterine growth restriction, cardiovascular disease, endothelial dysfunction, etc. Ultrasound is also the basis for the most widespread technology for fetal monitoring, that is, computerized cardiotocography, whose technical aspects and underlying physiological principles behind heart rate variability are presented from a biomedical engineering perspective, along with clinical applications.

From the other side, technologies based on the recording of electro-physiological signals from the baby represent a long-standing research topic that is progressively entering clinical practice. The principles behind fetal electrocardiography are presented along with the signal processing issues associated with non-invasive measurement of this elusive signal. On this basis, several innovative devices for multimodal fetal monitoring have been developed and are presented. Advanced signal processing methods can then be developed to analyze such signals for the identification of subtle patterns which predict sudden fetal deaths, as for the case of T-wave alternans. When something goes wrong during pregnancy or labor, the newborn is in a high-risk condition and admission to neonatal intensive care unit could be the last chance to save the baby. Advanced engineering and mathematical methods that could potentially provide effective assisting technology in this context are also presented.

Finally, a completely different perspective is introduced in the last chapter of the book, which opens to new frontiers metabolomics and perinatal programming, that is, the response by the developing organism to a specific challenge altering the trajectory of development with resulting persistent effects on phenotype.

Cagliari, Italy   Danilo Pani
Eindhoven, The Netherlands   Chiara Rabotti
Milan, Italy   Maria Gabriella Signorini
Ancona, Italy   Laura Burattini

Contents

Chapter 1
Ultrasound in Pregnancy –
From Ultrasound Physics to Morphological
and Functional Measurements of the Fetus

Massimo Mischi and Judith van Laar

Contents

M. Mischi (✉)
Eindhoven University of Technology (Netherlands), Electrical Engineering Department,
Biomedical Diagnostics Lab, Eindhoven, The Netherlands
e-mail: m.mischi@tue.nl

J. van Laar
Máxima Medical Center Veldhoven (Netherlands), Obstetrics and Gynaecology Department,
Veldhoven, The Netherlands

© Springer Nature Switzerland AG 2021
D. Pani et al. (eds.), *Innovative Technologies and Signal Processing in Perinatal Medicine*, https://doi.org/10.1007/978-3-030-54403-4_1

1.1 Ultrasound Physics

1.1.1 Ultrasound Generation

Ultrasound refers to pressure waves propagating through a medium. They are similar to audible sound waves, but at higher frequency beyond 20 kHz. In clinical diagnostic applications, frequencies between 1 and 10 MHz are typically employed. Ultrasound waves are generated by an ultrasound transducer, which makes the electromechanical conversion between pressure and electrical potential. Pressures from 10 to 1000 kPa are typically generated for diagnostic applications.

The electromechanical conversion is made by piezoelectric crystals. These can be viewed as charge generators. The generated electrical charge, q, is related to the mechanical force applied to the crystal surface, F, by the linear relationship $q = k_p F$, with k_p being the piezoelectric constant. The value of k_p ranges from 2.3 pC/N for quartz up to over 200 pC/N for PZT ceramics, which are typically employed in ultrasound transducers due to their efficient conversion.

Recalling the elastic law (or Hooke's law), and with reference to Fig. 1.1, the applied force F is a function of the crystal deflection, x, defined by $F = k_e x$, with k_e being the elasticity constant. We may therefore write $q = Kx$, with $K = k_e k_p$. The crystal can also be modeled as a plane capacitor, whose relationship between electrical potential, V, and electrical charge, q, is defined by the capacitance $C = q/V$. Therefore, combining all together, the relationship between electrical potential and crystal deflection reads as

$$V = \frac{Kx}{C}. \tag{1.1}$$

Accounting for electrical leakage through the crystal and capacitive effects, which are also introduced by the cables to the frontend amplifier, the transfer function in Eq. (1.1) shows frequency dependency. As a result, the electromechanical conversion can be described by a first-order system [1]. However, this frequency dependency is typically overcome by the employment of a charge amplifier, reestablishing the ideal (zero-order) transfer function in Eq. (1.1).

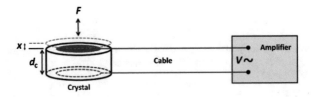

Fig. 1.1 Schematic representation of an ultrasound transducer made of a piezoelectric crystal of thickness d_c, performing the electromechanical conversion between electrical potential, V, and force normal to the crystal surface, F, which is related to the crystal deflection, x

Eventually, also accounting for the mechanical behavior (resonance) of the crystal, the full electromechanical conversion is well represented by a second-order system, showing resonance and anti-resonance frequencies. These frequencies provide full characterization of the ultrasound transducer. A simple approximated rule to determine the (mechanical) resonance frequency relies on the velocity, c_c, of ultrasound traveling through the crystal. For a crystal of thickness d_c, the resonance frequency f_0 can be derived as

$$f_0 = \frac{c_c}{2d_c},\tag{1.2}$$

i.e., the ultrasound wavelength through the crystal, $\lambda = c_c/f_0 = 2d_c$.

Most diagnostic applications require short ultrasound pulses of few oscillations. Therefore, epoxy baking is used in the transducer to dampen the crystal oscillations and permit the generation of short pulses.

The generated pulses are transmitted inside the body by positioning the ultrasound transducer on the skin. In order to optimize the transmission of the generated ultrasound waves into the body, it is important to reduce the mismatch in acoustic impedance between piezoelectric crystal and skin. To this end, the transducer surface is covered by a matching layer with thickness equal to $\lambda/4$ and *acoustic impedance*, Z_m, between that of the skin, Z_s, and the piezoelectric crystal, Z_c. The acoustic impedance, conventionally indicated with the symbol Z, provides the acoustic characterization of a medium; it is given as

$$Z = \rho_0 c_0,\tag{1.3}$$

with ρ_0 the medium density and c_0 the speed of sound. A typical value employed for the acoustic impedance of the matching layer is $Z_m = \sqrt{Z_s Z_c}$ [2].

1.1.2   Ultrasound Propagation

Assuming a linear medium, the ultrasound propagation velocity, c_0, can be approximated as $c_0 = \sqrt{B/\rho_0}$, with B [Pa] and ρ_0 [kg/m³] being the bulk modulus and density of the medium, respectively. Therefore, the propagation velocity of ultrasound differs for different media, with an average value of 1540 m/s in tissue. This results in typical values of the acoustic impedance, Z, ranging from 1.67 MRayls in blood to 7.9 MRayls in bone tissue. Notice that the acoustic impedance is measured in Rayls, with 1 Rayl = 1 Kg m^{-2} s^{-1}. Assuming the generated pressure wave to be well represented by a plane wave, its propagation can be described by the *wave equation* as

$$\frac{\partial^2 p(z,t)}{\partial z^2} = \frac{1}{c_0^2}\frac{\partial^2 p(z,t)}{\partial t^2}, \tag{1.4}$$

where z is the propagation axis, t is time, and p is the pressure amplitude. With reference to Fig. 1.2, the pressure p is related to the molecule displacement, u, induced by the oscillating surface of the transducer (deflection x), by the relation

$$p = \rho_0 c_0 \frac{\partial u}{\partial t} = \rho_0 c_0 v, \tag{1.5}$$

with v being the molecule velocity. Eq. (1.4) can simply be derived by the combination of the *momentum* and the *continuity* equations, neglecting second order terms and making use of the following linear relationship:

$$\rho(z,t) = \rho_0 + \frac{p(z,t)}{c_0^2}. \tag{1.6}$$

For a linear, lossless medium, a solution of Eq. (1.4) is given as

$$p(t,z) = p_0 e^{jk(c_0 t - z)}, \tag{1.7}$$

with $j = \sqrt{-1}$, p_0 the maximum pressure amplitude, and $k = \omega_0/c_0$ the *wave number*, where $\omega_0 = 2\pi f_0$. Usually the real part of the solution in Eq. (1.7), $\mathrm{Re}[u] = u_0 \cos(k(c_0 t - z))$, is used to represent ultrasound waves.

In the presence of viscous media, the acoustic pressure decays with the distance as the ultrasound energy is partly transformed into other forms, such as heat. This condition, referred to as absorption, can be represented by introducing a complex wave number $k = (\omega_0/c_0)\text{-}ja$. The solution of the wave equation is then given as [3].

$$p(t,z) = \mathrm{Re}\left[p_0 e^{j(\omega_0 t - kz)}\right] = p_0 \mathrm{Re}\left[e^{j\left(\omega_0 t - \left(\frac{\omega_0}{c_0} - ja\right)z\right)}\right] = e^{-az} p_0 \cos\left(\omega_0 t - \frac{\omega_0}{c_0}z\right). \tag{1.8}$$

The presence of the negative exponential, e^{-az}, represents the attenuation effect.

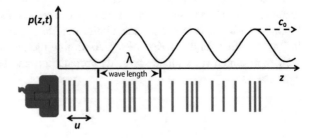

Fig. 1.2 Generation and propagation at velocity c_0 of an ultrasound wave, evidencing the pressure variation, $p(z,t)$, as a function of molecular displacement u. The wavelength, λ, is also indicated

In reality, the wave number also shows frequency dependency, which translates into frequency-dependent attenuation, $a(\omega)$, and speed of sound, $c(\omega)$. The latter is referred to as *dispersion*. These dependencies are defined by the *Kramers-Kronig relation* as

$$k = \frac{\omega}{c_0} + a(z) \mathrm{tg} \left[\frac{\pi n(z)}{2} \right] \omega |\omega|^{n(z)-1} - ja(z) |\omega|^{n(z)}, \qquad (1.9)$$

with n real positive number in [1, 2]. In addition to the energy loss due to absorption, also energy scattering contributes to a decay of the acoustic intensity. Altogether, these phenomena are taken into account by the *attenuation coefficient*, a, describing the exponential decay of the acoustic intensity over propagation distance, z, and frequency, $f = \omega/2\pi$, as

$$p(z) = p_0 e^{-azf^n}. \qquad (1.10)$$

The attenuation coefficient is typically represented in [dB cm^{-1} Hz^{-1}] by the coefficient $a_{dB} = 20\log_{10}(p/p_0)$. By simple exponential and logarithmic transformations, we can write

$$p(z) = p_0 e^{-0.115 a_{dB} z f^n}. \qquad (1.11)$$

It is clear that attenuation is proportional to the ultrasound frequency. This relation has important implications for ultrasound imaging. While higher resolution requires imaging at higher ultrasound frequency, leading to a shorter wavelength, λ, the achievement of good penetration for the investigation of deeper tissue requires imaging at lower frequency. The optimal compromise is application-dependent. While a transvaginal probe can operate at frequencies as high as 9 MHz, being in contact with the cervix and in close vicinity with the fetus, transabdominal ultrasound operates at lower frequencies of 2–4 MHz due to the higher distance from the fetus.

The energy carried by ultrasound waves is defined by the *acoustic intensity*, I, which represents the power across a unitary surface [W/m^2] at a certain distance z_0 from the transducer, and can be derived as

$$I(z_0) = \frac{1}{T} \int_T pv \, dt = \frac{1}{T} \int_T \frac{p^2(z_0,t)}{Z} \, dt = \int_T \cos^2 \left(2\pi f_0 t - kz_0 \right) dt = \frac{p_0^2}{2Z}, \qquad (1.12)$$

with f_0 being the ultrasound frequency, $p(z_0,t) = p_0 \cos(2\pi f_0 t - kz_0)$, and $T = 1/f_0$.

1.1.3   Ultrasound Echo

Ultrasound echo, which is the basis of echography, is the result of scattering and reflection. The medium where ultrasound propagates, in our case tissue, is made of a distribution of acoustic scatterers with dimension smaller than the wavelength λ. In typical clinical applications, λ is a fraction of a millimeter. Smaller scatterers, such as red blood cells (6–8 μm), will produce scattering of the incident ultrasound wave in all directions, which is referred to as *Rayleigh scattering*. For imaging purposes, we are mainly interested in the acoustic intensity that is scattered back to the ultrasound transducer, where the electromechanical conversion is performed. This scattering phenomenon is referred to as *backscattering*. The specific distribution of scatterers in tissue produces a combination of constructive and destructive interferences among the scattered waves, producing a typical texture in the reconstructed image, referred to as *speckle* (Fig. 1.3). The presence of speckle is a distinctive characteristic of ultrasound images with respect to other imaging modalities.

When the distribution of scatterers is organized according to a macroscopic geometry with size larger than λ, the backscattering translates into a reflection phenomenon, referred to as *acoustic echo*. Reflection occurs at the interface between different tissues and permits recognizing the boundary between different anatomical structures. By imposing the continuity of molecule displacement, u, and pressure amplitude, p, across the interface between tissues with different acoustic impedance, Z_1 and Z_2, the laws regulating reflection and transmission across an interface can be derived as

$$\frac{p_r}{p_i} = \left(\frac{Z_2 \cos\theta_i - Z_1 \cos\theta_t}{Z_2 \cos\theta_i + Z_1 \cos\theta_t} \right),$$
$$\frac{p_t}{p_i} = \left(\frac{2Z_2 \cos\theta_i}{Z_2 \cos\theta_i + Z_1 \cos\theta_t} \right),$$

$$(1.13)$$

with θ_i the incident angle and θ_t the transmission (refraction) angle [4]. The symbols p_i, p_r, and p_t represent the pressure amplitude of the incident, reflected, and transmitted wave, respectively (Fig. 1.4).

Fig. 1.3 Example of speckle pattern

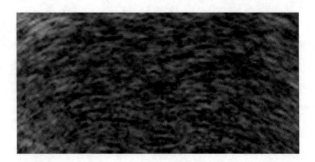

Fig. 1.4 Reflection and transmission of an ultrasound wave across a discontinuity in acoustic impedance (from Z_1 to Z_2). The incident, reflected, and transmitted pressure are indicated with the symbols p_i, p_r, and p_t, respectively

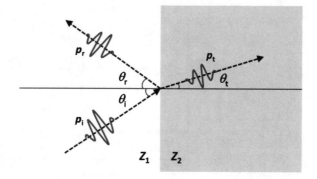

The relations in Eq. (1.13) can also be derived for the intensity, I, by using Eq. (1.12). The angles between incident, reflected, and transmitted wave are simply related according to *Snell's law* as

$$\frac{\cos\theta_i}{c_1} = \frac{\cos\theta_r}{c_1} = \frac{\cos\theta_t}{c_2},\tag{1.14}$$

with θ_r representing the reflection angle, and with c_1 and c_2 being the speed of sound in tissue 1 and 2, respectively.

1.2   Ultrasound Imaging Technology

1.2.1   Ultrasound Beam Profile

Echography is based on the analysis of echoes resulting from the reflection of ultrasound short pulses (usually few wave cycles) that are transmitted inside the body. As from Eq. (1.13), discontinuities in acoustic impedance, Z, produce echoes at the interface between different tissues or organs. Relevant information results also from the backscatter generated within the same tissue, which produces a deterministic speckle pattern that is specific for a certain tissue structure (Fig. 1.3).

The time delay, Δt, between a transmission event and the arrival time of the received echoes provides information about the depth, d, where echoes are generated (Fig. 1.5). This is based on the assumption of a fixed speed of sound $c_0 = 1540$ m/s, such that $d = c_0\Delta t/2$. The *longitudinal resolution* is determined by the pulse envelope as $N\lambda/2$, with N being the number of cycles in a pulse and $\lambda = c_0/f_0$ the wavelength for a chosen ultrasound frequency f_0.

The *lateral resolution* is related to the ultrasound beam profile, being a measure or the narrowness of the ultrasound beam. It can be derived by applying the *Huygens principle*, integrating over infinite point sources distributed over the emitting surface of the transducer (S in Fig. 1.6). Before applying the Huygens principle, the

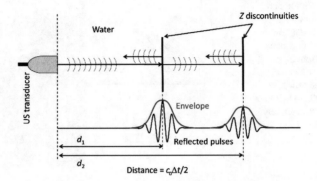

Fig. 1.5 Schematic representation of two echoes produced by two Z (acoustic impedance) discontinuities in water. The distance, d, of the discontinuities from the transducer is related to the delay at which the echoes are received. The envelope of the reflected pulses is also shown, which is used to represent the echo signal and determines the longitudinal resolution

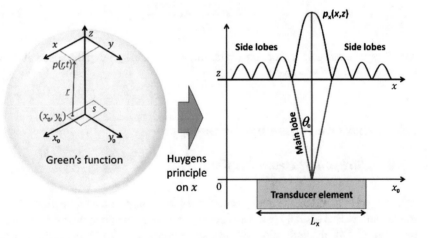

Fig. 1.6 On the left, application of the Green's function to calculate the pressure at point $\underline{r}$, $p(\underline{r}, t)$, generated by a source point on the emitting surface S. By integration of all contribution from the emitting surface S, based on the Huygens principle, the pressure field can be determined. On the right, integrating on the aperture (L_x) only, the pressure field at distance z from the transducer, $p_x(x,z)$, is determined along the lateral direction x. In the far field, this is approximated by a sinc(.) function, showing a main lobe of angle θ_0 surrounded by side lobes

ultrasound pressure field generated by a point source should be determined. To this end, we derive the Green's function $G(\underline{r},t)$ in space, $\underline{r} = (x,y,z)$, and time, t, which solves the 3D wave equation as

$$\nabla^2 G(\underline{r},t) - \frac{1}{c_0^2}\frac{\partial G(\underline{r},t)}{\partial t} = -\delta(\underline{r})\delta(t),$$

$$\text{with } G(\underline{r},t) = \frac{\delta\left(t - \frac{|\underline{r}|}{c_0}\right)}{4\pi|\underline{r}|}.$$

(1.15)

In Eq. (1.15), $\delta(\underline{r})$ and $\delta(t)$ represent the Dirac impulse in space and time, respectively. Therefore, the Green's function can simply be interpreted as the spatiotemporal impulse response of the system, and the pressure field generated by a point source producing harmonic pressure variations at frequency f_0 can be derived as

$$p(\underline{r},t) = G(\underline{r},t) *_{\underline{r},t} \delta(\underline{r})\cos(2\pi f_0 t),$$

(1.16)

with the symbol $*$ being the convolution operator. A solution for a homogeneous medium is given as

$$p(\underline{r},t) = Re\left[\frac{p_0 e^{2\pi f_0\left(t - \frac{|\underline{r}|}{c_0}\right)}}{4\pi|\underline{r}|}\right] = \frac{p_0\cos(2\pi f_0 t - k|\underline{r}|)}{4\pi|\underline{r}|}.$$

(1.17)

The pressure field generated by an emitting surface (transducer element) can be derived by the *Rayleigh-Sommerfeld integral*. As shown in Fig. 1.6, this corresponds to the application of the Huygens principle integrating over a distribution of point sources (x_0, y_0) over the emitting surface. The contribution of each point source is described as in Eq. (1.17). For a detailed derivation of the integral solution for a rectangular emitting surface, the reader may refer to [5].

Assuming a large distance from the emitting surface, *Fraunhofer* approximation for the so-called *far field* ($\lambda z \gg x^2$ and $\lambda z \gg x_0^2$), the pressure field can be approximated as the 2D Fourier inverse transform of the emitting geometry. For a rectangular geometry, the Fourier integral can be separated in the two directions, x_0 and y_0, referred to as *aperture* and *elevation*. With focus on the aperture, represented by the rectangle $\prod_{L_x}(x_0)$, and neglecting additional multiplicative factors, we can write

$$P_x(x,z) = \int_{-\infty}^{\infty} \prod_{L_x}(x_0) e^{j2\pi\left(\frac{x}{\lambda z}\right)x_0} dx_0 = L_x \text{sinc}\left(\frac{\pi x L_x}{\lambda z}\right),$$

(1.18)

with L_x being the transducer aperture (length in the x_0 direction) and z the distance from the transducer (depth). The Fourier integral in Eq. (1.18) formulates the transformation from the variable x_0 to $x/\lambda z$. In the far field, x/z can be approximated by $\sin \theta_0$, where θ_0 indicates the aperture angle with respect to the longitudinal axis z. The first zero-crossing of the sinc(.) function in Eq. (1.18), given for $x/z = \lambda/L_x$, provides the size of the ultrasound main lobe, which increase for increasing depth as described by the angle $\theta_0 = \sin^{-1}(\lambda/L_x)$. Often the −6 dB full width of the lobe, typically referred to as *full-width half-maximum* (FWHM), is derived from Eq. (1.18) and used to derive the lateral resolution as

$$\text{FWHM} = 1.206 \frac{\lambda z}{L_x}. \tag{1.19}$$

It is clear that an increase in the central frequency f_0 of the ultrasonic pulses, resulting in shortening of the wavelength $\lambda = c_0/f_0$, produces an increase in both lateral and axial resolution. However, higher frequencies are also subjected to higher attenuation. Improved lateral resolution can also be obtained by extending the aperture of the transducer. Unfortunately, many applications, such as transvaginal ultrasound, pose important constraints to the transducer size, which must be kept limited.

Until now we have discussed only on the main lobe described by the first zero-crossing of the sinc(.) function in Eq. (1.18). However, the additional *side lobes*, described by the following zero-crossings for $x/z > \lambda/L_x$, should also be taken into account as they may introduce image artifacts from directions that are different from the main lobe. Reduction in the side lobes can be obtained by *apodization*, i.e., by replacing the rectangular function in Eq. (1.18) with a window function, such as Hamming or Hanning [6], which is designed to reduce the side lobes at the cost of a lower lateral resolution. This approach is similar to the windowing approach used in time series, but then in space domain.

The provided description of the beam profile is valid in the so-called far-field approximation, i.e., for large distance from the transducer. As shown in Fig. 1.7, two different zones can be distinguished depending on the distance from the transducer, namely, the *Fresnel* and *Fraunhofer* zone. For a circular transducer of radius R, the

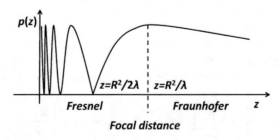

Fig. 1.7 Pressure field (maximum pressure p_0) produced by a circular transducer of radius R as a function of distance z from the transducer. The last zero and the focal distance are also indicated, along with near (Fresnel) and far (Fraunhofer) zone

transition between the Fresnel and Fraunhofer zone occurs at distance $z = R^2/\lambda$ from the transducer. This distance also corresponds to the maximal lateral resolution (approximately equal to R) and natural focus of the beam, showing the maximum amplitude because of the constructive interference of the emitted waves. As the depth increases, in the Fraunhofer zone, the lateral resolution decreases.

1.2.2 Array Beam Forming

Up until now we have described the pressure field generated by a single element. However, in order to create images, arrays of multiple elements are typically employed. In an array transducer, the emitting surface of aperture L_x is made of a finite number of rectangular elements adjacent to each other, each with aperture w. The distance d between the centers of the adjacent elements is constant and is referred to as *pitch*. The emitting surface can thus be described as a train of Dirac impulses of period d, convolved with a rectangular function of size w. The pressure field can then be derived according to Eq. (1.18) by application of the inverse Fourier transform, F (.), as

$$
\begin{aligned}
p_x\left(x,z\right) = \mathcal{F}^{-1}\left\{\left[\prod_w[x_0]*\sum_m \delta\left(x_0 - md\right)\right]\prod_{L_x}[x_0]\right\} = \\
= \mathcal{F}^{-1}\left\{\prod_w[x_0]\right\}\mathcal{F}^{-1}\left\{\prod_{L_x}[x_0]\sum_m \delta\left(x_0 - md\right)\right\},
\end{aligned}
\tag{1.20}
$$

with m the element index in the array and x_0 representing the x direction on the array transducer surface. By application of the sampling theorem, Eq. (1.20) yields

$$
p_x\left(x,z\right) = \frac{L_x w}{d}\,\mathrm{sinc}\left[\frac{\pi wx}{\lambda z}\right]\sum_m \mathrm{sinc}\left[\frac{\pi L_x}{\lambda}\left(\frac{x}{z} - \frac{m\lambda}{d}\right)\right],
\tag{1.21}
$$

which represents the pressure field along the x direction at distance z from the transducer array. With the approximation $x/z = u = \sin(\theta)$, Eq. (1.21) can also be interpreted as the pressure field for varying azimuth angle θ.

Two main components characterize the derived pressure field. First, the profile given by each individual element, $\mathrm{sinc}(\pi w\lambda u)$, which introduces a smooth weighting function over the full pressure profile because of the small element size, w. Secondly, the train of functions, $\mathrm{sinc}[\pi L_x(u-m\lambda/d)/\lambda]$, caused by the full array of length L_x sampled at the spatial period d (pitch). These spectral replicates, resulting from the application of the sampling theorem, introduce artifacts in the image which are referred to as *grating lobes* (Fig. 1.8). In fact, at angles $\theta_g = \sin^{-1}(m\lambda/d)$, the main lobe along with the side lobes is repeated, investigating directions in the field of

Fig. 1.8 Schematic representation of the pressure field produced by a linear array transducer (aperture = L_x and pitch = d) along the lateral direction. Because of spatial sampling, the sinc(.) function in Fig. 1.6 is repeated multiple times, forming the so-called grating lobes at angles that are multiple of θ_g

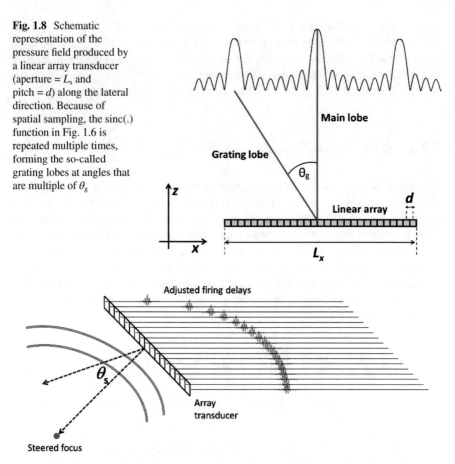

Fig. 1.9 Schematic representation of electronic steering and focusing by adjusting the delay between the pulses fired from different channels. The reader should notice that this procedure can be implemented both in transmit and receive

view that are different from the intended main lobe. To avoid grating lobes in the field of view, the condition $d < \lambda$ must be met.

The possibility of driving independently the individual array elements facilitates the implementation of apodization functions for reduction of the side lobes by using different weights for the signals driving different elements. It also permits electronic steering the beam in different directions, as well as electronic focusing at distances that differ from the natural focus of the array, as shown in Fig. 1.9.

These features are implemented by *delay-and-sum* operations, adjusting the delays between signals corresponding to different transducer elements. In particular, steering by an angle θ_s is obtained by introducing delays Δt between adjacent elements such that

$$\theta_s = \sin^{-1}\left(\frac{c_0 \Delta t}{d}\right). \tag{1.22}$$

Beam steering is also reflected on the grating lobes, which will be steered by the same steering angle θ_s. To avoid grading lobes appearing in the field of view up to a maximum (unrealistic) steering angle $\theta_s = 90$ degrees, the array pitch must meet the condition $d < \lambda/2$. Both electronic focusing and beam steering can be implemented in transmit and/or receive.

In fact, up until now we have discussed the amplitude of the transmit pressure profile only. The same description applies to the receive profile. Eventually, an echographic image results from the convolution of the full impulse response of the imaging system (transmit and receive) with the impulse response characterizing tissue backscatter and attenuation. This impulse response should also account for the pressure variations in time domain, which depend on the transmitted ultrasound pulse [5].

Following a similar derivation as for Eq. (1.18), the full spatiotemporal response of an ultrasound imaging system, accounting for the aperture direction only, can be described as

$$p(x,z,t) = \rho_0 h_x(x,z,t) * \frac{\partial v(x_0,t)}{\partial t},$$

$$h_x(x,z,t) = \frac{c_0}{x}\sqrt{\frac{z}{2\pi}} \prod_{xL_x/zc_0}(t), \tag{1.23}$$

$$v(x_0,t) = v_0 e^{j\omega(t-t_0)} e^{-\frac{\sigma_\omega^2(t-t_0)^2}{2}}.$$

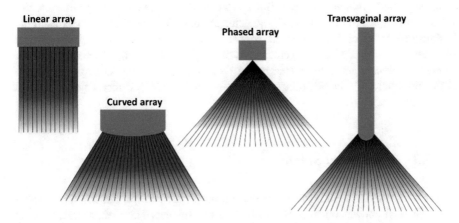

Fig. 1.10 Different types of array transducers and corresponding fields of view

In Eq. (1.23), $h_x(x,z,t)$ represents the full spatiotemporal impulse response of the ultrasound imaging system along the aperture, with $v(x_0,t)$ the displacement velocity of the emitting surface of the transducer element, typically corresponding to a Gaussian pulse of amplitude v_0 and bandwidth σ_ω. To obtain the full response of the system in 3D, an additional convolution with the impulse response along the elevation direction, $h_y(y,z,t)$, should be included.

Ultrasound imaging is realized by covering a field of view with different scan lines (Fig. 1.10), determined by the main lobe. Depending on the array size and shape, these lines can be realized by activating adjacent subsets of elements along the array or by steering the beam in different directions (phased array). For each line, one pulse is transmitted and the resulting echoes received back. In order not to confuse reflections coming from different pulses, one pulse is transmitted only after all reflections from the previous pulse have been received, at least up to the maximum depth that should be investigated. Therefore, the frame rate of ultrasound imaging is inherently limited by the speed of sound. As an example, imaging a frame made of 200 lines that are 10-cm deep will take 26 ms, leading to a frame rate of about 38 Hz.

Nowadays, enabled by increased computational power, ultrafast ultrasound imaging solutions are becoming available that allow increasing the frame rate up to several thousands of frames per second, beyond any other imaging modality [7]. To this end, *plane* or *diverging waves* are transmitted in order to insonify the full field of view. The image is then reconstructed in receive only, typically resulting in poor quality. In order to improve the image quality, several transmissions can be performed at different steering angles, and the resulting echo images combined. This approach is referred to as spatial (coherent) compounding [8]. As a result, the frame rate is reduced by the number of transmissions that are used for compounding.

Compounding can also be used in standard focused imaging for further improving the image quality and reducing speckle noise. Not only spatial compounding, but also frequency compounding can be used for improving the image quality. Improvement of the image quality can also be obtained by the implementation of the so-called *synthetic aperture* [9]. Instead of firing and receiving with each element (or subset of elements), all elements are used to receive the echoes from the transmission of each separate element in sequence. All the resulting low-quality images are then combined together to form a high-quality image. Depending on the exact implementation, the frame rate is similar to that of standard focused beamforming, but with the advantage of focusing in transmit and receive over the full field of view.

1.2.3 Ultrasound System

The ultrasound system implements the full imaging chain from ultrasound electro-mechanical conversion to image formation and display. The main elements of an ultrasound system are the following:

- The ultrasound transducer, executing the electromechanical conversion.
- The frontend that interfaces the transducer, comprising the transmit/receive amplifiers and switches, along with the analog-to-digital (and vice versa) converters. The receive amplifiers implement *time gain compensation* (TGC), i.e., the received signals may undergo varying amplification depending on their time delay, corresponding to the depth where the received echoes are originated. This allows compensating for ultrasound attenuation, described by Eq. (1.10).
- The scanner, comprising the transmit/receive beamformers along with the master clock timing the full system.
- The backend, comprising the core processing unit that controls the image formation chain, along with the digital scan converter and post-processing units for image analysis, enhancement, compression, and display.
- The user interface, comprising display and keyboard, along with all the controllers for adjusting the image acquisition parameters and for optimizing the image quality. A number of predefined settings are usually available to the user.

Several steps are taken in order to generate an image, and the sonographer is given the opportunity to optimize the image quality by adjusting a number of parameters through specific controllers to be properly set. This requires experience and understanding of the imaging chain. The main steps leading to the formation of an ultrasound image are summarized hereafter.

The transducer elements convert the received echoes in electrical signals, which are amplified according to the gain and TGC controllers. Besides the receive amplifiers, controlled by gain and TGC, the user can also set the pressure amplitude of the transmitted pulses, which also determines the amplitude of the received echoes. There are however limitations to the amount of acoustic intensity (see Eq. (1.12)) that can be delivered to tissue, determined by safety guidelines [10].

The amplified signals form scan lines through the delay-and-sum operation. The selected elements and delays determine the scan line direction and focus (Fig. 1.9). The latter, as well as the spanned field of view, can be determined by the user. Each line investigates a different direction in the field of view. On each line,

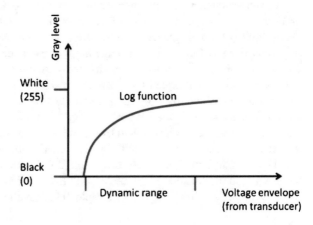

Fig. 1.11 Logarithmic compression mapping the dynamic range onto 256 gray levels

demodulation is applied in order to extract the signal envelope, whose amplitude relates to the echo intensity (Fig. 1.5). As shown in Fig. 1.11, *log-like compression* is then applied to the demodulated lines in order to map the dynamic range of the signal onto a range that is quantized in 256 gray levels, coded by 8 bits rendered by the display [11].

For proper image visualization, it is important to set the dynamic range such that the resulting image is well distributed over the 256 gray levels, not too dark and not too bright (signal saturation) [11]. Additional refining can be applied in post-processing, by the employment of nonlinear maps that allow improving contrast in the required intensity range.

Digital images, either for display or for storage, are based on a Cartesian grid of pixels. Therefore, especially for diverging scan lines (due to either beam steering or probe convexity, as shown in Fig. 1.10), these must be interpolated in order to form a regular Cartesian grid. This operation, referred to as scan conversion, also determines the image resolution. Especially for diverging scan lines originating from convex probes, scan conversion dominates the lateral resolution over the beam profile described in Eq. (1.18).

The formed image, as discussed in Sect. 1.1.3, presents a typical speckle texture which results from the interference of the backscattered ultrasound waves (Fig. 1.3). The statistics of the speckle amplitude is well described by a Rayleigh probability density function (PDF) as

$$\text{PDF}(A) = \frac{A}{\sigma^2} e^{-\frac{A^2}{2\sigma^2}}, \tag{1.24}$$

with A being the signal envelope and σ the mean scattering strength [12]. In spite of the statistical description of its amplitude, the speckle presents a deterministic pattern that depends on the spatial distribution of the (tissue) scatterers and on the adopted ultrasound frequency. This pattern moves along with tissue, giving the opportunity to image and analyze tissue deformation and strain by speckle-tracking techniques [13].

The fundamental modes in which an ultrasound system is used are distinguished in *A-mode*, *M-mode*, and *B-mode*, which define the generation of a single scan line, the dynamic (repeated) generation of the same scan line, and the generation of multiple scan lines to form a (gray-level) image, respectively. M-mode ultrasound is especially indicated for fast-moving structures, such us cardiac structures, which must be investigated at higher rate (temporal resolution). Nowadays, B-mode ultrasound is also extended to 3D by using either electromechanical wobbling probes or electronic steering with matrix probes (Fig. 1.12). Although the achieved frame rates and resolutions are lower than by standard B-mode ultrasound, 3D visualization may provide a valuable aid for the identification of fetal abnormalities. Moreover, spatio-temporal image correlation (STIC) post-processing allows for the reconstruction of ultrasound volume loops at B-mode frame rates by acquisition of a 2D sweep or "stitching" together multiple 3D acquisitions [14]. In the future, the

Fig. 1.12 Transabdominal
3D ultrasound image
showing the face of a fetus.
Details of the facial
morphology are
clearly visible

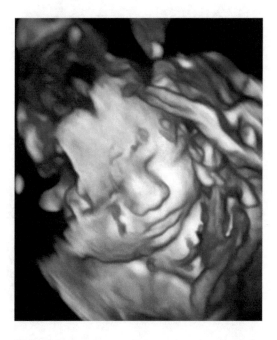

development of ultrafast imaging presented in Sect. 1.2.2 is expected to provide new
solutions to achieve 3D real-time imaging of the fetus at high volume rates.

1.3 Ultrasound Doppler Technology

1.3.1 Doppler Principle

Doppler ultrasound is among the fundamental operational modes of an ultrasound
system, next to A-mode, M-mode, and B-mode ultrasound. It is based on the
Doppler effect, a frequency shift that is induced in the received signal by moving
scatterers. This principle can be used to estimate blood velocity as shown in
Fig. 1.13.

A transmitter (Tx) and a receiver (Rx) are integrated in the same probe. The
intersection between the transmit and receive beams forms the *sample volume*,
which contributes to the received signal. Blood flow velocity in the sample volume
produces a shift in the received ultrasound frequency. This is induced by blood cells
in motion (flow), which act as moving scatterers. The focal depth, determining the
position of the sample volume, is determined by the angle β, which depends on the
transducer geometry and, in some systems, can be mechanically adjusted.

A continuous ultrasound wave is transmitted by the Tx element at frequency f_t.
The moving scatterers introduce a frequency shift in the reflected wave, such that
the received frequency, f_r, at the receiver Rx is given as

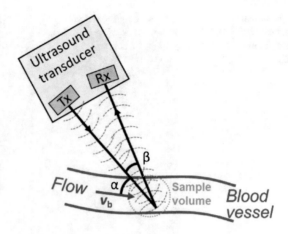

Fig. 1.13 Use of the ultrasound principle for measuring blood velocity. A transmitter (Tx) and a receiver (Rx) are used to transmit and receive ultrasound. The angle α determines the angle between blood velocity (v_b) and the transmitted ultrasound beam, while the angle β determines the angle between the transmitted and received beams. The received signal integrates all the backscatter occurring in the sample volume

$$f_r = f_t \left(\frac{c_0 - v_b \cos\alpha}{c_0 + v_b \cos(\alpha + \beta)} \right), \tag{1.25}$$

with v_b being the blood velocity and c_0 the speed of sound. The angle α represents the angle between ultrasound beam and blood velocity. For small β and $v_b \ll c_0$, the frequency shift $\Delta f = f_t - f_r$, known as the Doppler frequency, can be approximated as

$$\Delta f = f_t \left(1 - \frac{c_0 - v_b \cos\alpha}{c_0 + v_b \cos(\alpha + \beta)} \right) \cong v_b \left(\frac{2 f_t}{c_0} \right) \cos\alpha, \tag{1.26}$$

Therefore, since the transmit frequency, f_t, and ultrasound speed, c_0, are known, the Doppler frequency Δf is directly proportional to the blood velocity, v_b. Interestingly, one can notice that the Doppler frequency for physiological blood velocities falls within the human audibility range. Therefore, a loudspeaker is typically used as interface of a Doppler system; gynecologists and sonographers are trained to interpret the produced sound tones in relation to blood velocity and turbulence.

Equation (1.26) clearly shows the dependency between Doppler frequency and incident angle α. This makes estimation of the absolute velocity restricted to a few applications where α is known. In fact, such controlled condition does not easily occur in pregnancy monitoring.

1.3.2   Receiver Architecture

To calculate Δf, the received signal is demodulated by a mixer that multiplies it by the transmitted signal. With $S_t = a_t\cos(2\pi f_t t)$ being the transmitted signal and $S_r = a_r\cos(2\pi f_r t)$ the received signal, the product of S_t by S_r gives $S_t \cdot S_r = a_t a_r[\cos(2\pi (f_t - f_r)t) + \cos(2\pi(f_t + f_r)t)]/2$. After suppressing the high-frequency component $(f_t + f_r)$ by low-pass filtering, the remaining signal is the Doppler signal at frequency $\Delta f = f_t - f_r$.

The lower frequencies in the Doppler signal typically represent noise introduced by slow tissue motion rather than blood flow. These low-frequency components, which deteriorate the Doppler estimates, are referred to as *clutter*. Clutter may for instance be caused by the cyclic pulsation of vessel walls which are included in the sample volume. In order to suppress clutter noise from the received signals, a high-pass filter, referred to as *clutter filter* (or *wall filter*), is applied to the signals.

A limitation of the presented demodulation strategy resides in the inability to distinguish forward from reverse blood flow. To make this distinction feasible, a *quadrature demodulation* architecture is used instead. This allows separating the in-phase, S_I, from the quadrature, S_Q, component of the received signal by the employment of two mixers multiplying the received signal by the transmitted signal and by a 90-degree phase-shifted version of it. From the generated S_Q and S_I components, a complex *Discrete Fourier Transform* (DFT) of the received echoes can be computed. The resulting frequency spectrum may be asymmetric, resulting in separable positive and negative frequency shifts. This output signal is usually visualized by a time-frequency representation, referred to as *spectral Doppler*. Alternatively, a two-channel filter can be employed to separate measurements of reverse (toward the probe) and forward blood (away from the probe) flow velocities. These two components, in the audio range, can be directed to the left and right channels of an audio headset.

1.3.3   Continuous and Pulsed Doppler

The system described until now is referred to as *Continuous Wave* (CW) *Doppler*. Ultrasound is continuously transmitted and all the signal backscattered in the focal region (sample volume) is continuously received and analyzed.

CW Doppler is however limited by the fact that it does not allow discriminating different moving scatterers, but integrates all the signal originating from the sample volume, which is determined by the focus. This limitation is overcome by *Pulsed Wave* (PW) *Doppler*, typically performed with standard array transducers. Each element of the transducer is used both as a transmitter and a receiver, similar to ultrasound imaging; a time sequence of short pulses is transmitted and received at the maximum *pulse repetition frequency* (PRF). The maximum PRF is determined by the investigated depth. It is therefore possible to define a specific target area where

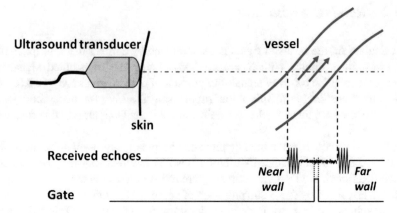

Fig. 1.14 Schematic representation of a pulsed wave (PW) Doppler system with the positioning of a gate between the near and far wall of a vessel

the Doppler shift is estimated. To this end, a time window corresponding to a depth interval of interest, referred to as *gate*, is positioned to extract the segment of received signal where the Doppler shift is estimated (Fig. 1.14). This can also be performed manually by positioning a so-called "caliper" on the image. This procedure can be repeated automatically for multiple depth intervals and the decomposed signals processed concurrently into different channels. This technique, referred to as *multi-gate pulsed Doppler*, enables reconstructing the velocity profile by combination of the estimated Doppler shift from each channel.

Similarly to standard ultrasound imaging, the achievement of a high longitudinal resolution with *PW Doppler* requires the transmission of short pulses. However, together with the pulse duration, also the gate duration determines the longitudinal resolution, which translates into the length of the sample volume. The lateral resolution, which translated into the width of the sample volume, is determined by the beam profile as described in Sect. 1.2.1. In general, the sample volume of a PW Doppler system is much smaller than that of a CW Doppler system.

Figure 1.15 shows a representation of how a PW Doppler system works. A train of pulses is transmitted at maximum PRF, given by $c_0/2d$, where d is the maximum depth to be investigated. As the target scatterer moves between two pulses by Δd ($\Delta d = v_b/\text{PRF}$), the subsequent received pulse (from a certain gate) is time-shifted by $2\Delta d/c_0$. The received pulses are represented in the so-called *fast-time* domain. As the full system is synchronized at the PRF, only one sample is taken from each pulse. The collection of all these samples forms the output signal, represented in the so-called *slow-time* domain. It can be easily proven that the frequency content of the output signal reconstructed in the slow-time domain corresponds to the frequency shift derived with Eq. (1.26). It is therefore equivalent to the Doppler signal.

Compared to CW Doppler, PW Doppler comprises a sampling process, which may therefore introduce *aliasing* effects. In order to avoid this, the Nyquist relation PRF > $2\Delta f$ must be satisfied, where Δf is the peak Doppler frequency associated

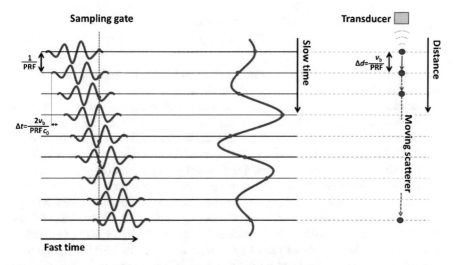

Fig. 1.15 Schematic representation of a pulsed Doppler system. Pulses are fired at the pulse repetition frequency (PRF) and their echoes (fast time) sampled at the PRF. With the scatterer moving at velocity v_b away from the transducer, each received echo will show a delay equal to $2v_b/\mathrm{PRF}c_0$, corresponding to the displacement of the scatterer by a distance $\Delta d = v_b/\mathrm{PRF}$. The resulting sampled signal in the slow time shows a frequency corresponding to the Doppler shift

with the peak blood velocity, v_p. Recalling Eq. (1.26), the Nyquist limit translates into the following condition for the maximum observable velocity without aliasing:

$$v_p = \mathrm{PRF}\left(\frac{c}{4f_t\cos\alpha}\right) = \frac{c^2}{8f_t d\cos\alpha}. \tag{1.27}$$

Equation (1.27) shows that for a shorter distance d, and hence a higher PRF, the maximum detectable velocity without aliasing is higher. Lower transmit frequency, f_t, also results in higher v_p. However, with reference to Eq. (1.26), lower f_t reduces the sensitivity of the Doppler system.

1.3.4   Velocity Estimators

In CW Doppler, the velocity estimation is directly derived by Eq. (1.26) from the Doppler frequency spectrum, Δf, estimated after quadrature demodulation. The same approach can be taken with PW Doppler based on the slow-time domain. As already discussed, a time-frequency representation of the signal (spectral Doppler) can be displayed that represents the velocity distribution in the sample volume.

PW Doppler signals can also be processed in the fast-time domain; the delay between subsequent reflected pulses, which can be determined by the peak of their cross-correlation, is related to the blood velocity, v_b, as

$$\Delta t = \frac{2v_b \cos\alpha}{c_0 \text{PRF}}.$$ (1.28)

As c_0 and the PRF are known, the estimation of Δt by cross-correlation allows esti-mating the blood velocity, v_b, by Eq. (1.28) [15].

Because of spectral broadening, the estimated Doppler spectrum is not concen-trated in a single spectral line, but is rather distributed over several frequencies, corresponding to a velocity distribution. Several factors contribute to spectral broad-ening. An important contribution to spectral broadening derives from the size of the sample volume, which integrates different velocity amplitudes (in the presence of velocity gradients) and directions (in the presence of turbulence). An additional contribution is given by velocity accelerations or decelerations during the time seg-ment (ensemble) adopted to estimate the Doppler spectrum. Geometric spectral broadening also occurs which is determined by the size and geometry of the trans-ducer element. Application of the Huygens's principle results in the integration of Doppler shifts that derive from a distribution of incidence angles, α, which show slight variation depending on the size of the transducer.

Although spectral broadening also carries diagnostic information, e.g., in rela-tion to turbulence, in several applications the estimation of a single value of velocity is preferred. To this end, dedicated strategies have been developed to estimate the *mean frequency* of the Doppler spectrum, Δf_m, which translates into mean velocity by Eq. (1.26). As proposed by Kasai et al. [16], Δf_m corresponds to the first statisti-cal moment of the Doppler power spectrum, $S_{ff}(f)$, which can be estimated by the autocorrelation function of the Doppler signal, $R(t)$, and its derivative, $R'(t)$, calcu-lated at time $t = 0$. Therefore, Δf_m can be derived as

$$\Delta f_m = \frac{\int_{-\infty}^{\infty} f S_{ff}(f) df}{\int_{-\infty}^{\infty} S_{ff}(f) df} = \frac{1}{j} \frac{R'(0)}{R(0)},$$ (1.29)

Loupass et al. proposed an implementation of Kasai's autocorrelator based on the in-phase and quadrature components, S_I and S_Q [17]. Given an ensemble of N sub-sequent samples from a certain depth, Δf_m can be estimated as

$$\Delta f_m = \frac{\text{PRF}}{2\pi} \text{tg}^{-1} \left(\frac{\sum_{n=0}^{N-2} S_Q(n+1) S_I(n) - S_I(n+1) S_Q(n)}{\sum_{n=0}^{N-2} S_I(n+1) S_I(n) + S_Q(n+1) S_Q(n)} \right).$$ (1.30)

1.3.5   Color and Power Doppler

The PW Doppler technique can be generalized in 2D by applying multi-gate PW along multiple scan lines generated with an array transducer. Obviously, the resulting PRF will decrease by the number of scan lines. This solution allows investigating an entire region of interest. The estimated velocities are mapped into a color-coded image; therefore, this Doppler imaging technique is referred to as *color Doppler*. Forward and reverse flows are represented in red and blue colorization, respectively. The absolute velocity is then given by the color luminance. The color coding is displayed in a color bar next to the color Doppler image (Fig. 1.16).

The color Doppler image is typically superimposed on the B-mode image. This solution, referred to as *duplex* scanner, allows simultaneous visualization and investigation of morphology and flow, aiding the sonographer with positioning the transducer and with the interpretation of the images. Unfortunately, a duplex scan also results in further reduction of the frame rate. The lower PRF may result in the presence of aliasing, which shows as abrupt transition from red to blue in the color maps. When higher frame rates are required, the region for the estimation of color Doppler can be limited to a smaller box, referred to as *Doppler box*, defined on the display.

Low sensitivity and high noise levels make color Doppler unsuitable to visualize and investigate low blood flow and perfusion. *Power Doppler* imaging was developed to circumvent this problem [18]. Different from color Doppler, where the Doppler frequency shift, Δf, is color coded, with Power Doppler the color scale

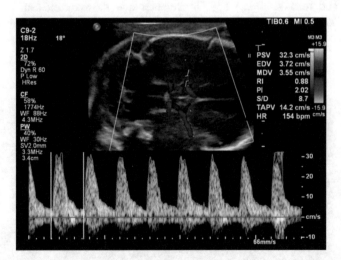

Fig. 1.16 Color Doppler image of the middle cerebral artery of a fetus in the second trimester. A box is visible that defines a specific region of interest to derive and display the color Doppler image, overlaid over the B-mode image (duplex scanner). A gate is also positioned and displayed on the image where spectral Doppler is estimated. The time-frequency representation of the estimated spectral Doppler is displayed on the bottom

relates to the total power of the Doppler signal. By integrating the contribution of all the Doppler frequency components, power Doppler shows increased sensitivity to low velocities. Moreover, power Doppler is not subject to aliasing artifacts. On the other hand, any information on the flow direction is lost. Power Doppler is especially useful for analyzing low flow in small vessels and tissue perfusion, where the blood volume fraction is low. Typical applications in pregnancy monitoring relate to the assessment of fetal brain perfusion.

1.4 Ultrasound in Pregnancy

1.4.1 First Trimester Scans

The Yolk sac, with a size of about 2 mm, is the first structure that is visible with ultrasound at five-week gestation. This is an extra-embryonic membrane within the chorionic cavity that plays a role in early embryonic blood supply. Next to the Yolk sac, an echogenic area is also visible which corresponds to the embryo (Fig. 1.17). At 6–7 week gestation, the embryo shows a growth in size to over 4 mm [19]. The size of the embryo is determined by the crown-rump length (CRL), as shown in Fig. 1.18. At this stage, also the amniotic sac becomes visible on ultrasound as a thin membrane. The inner of this membrane, referred to as amnion, encloses the amniotic cavity, containing the amniotic fluid and the embryo. The outer membrane, the chorion, forms the chorionic cavity that contains both amnion (surrounding the embryo) and yolk sac (Fig. 1.17). On the outer side, the amniotic sac is connected to the yolk sac. Irregularities in the shape and size of these structures may suggest a higher risk of abnormalities.

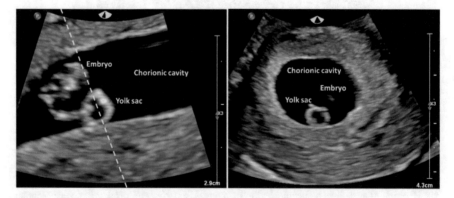

Fig. 1.17 Transvaginal ultrasound images showing a longitudinal (left side) and transversal (right side) view of a chorionic cavity including both yolk sac and embryo. The dashed line on the longitudinal view represents the plane shown in the transversal view

Fig. 1.18 Crown-rump length (CRL) measurement in an embryo of 7 weeks

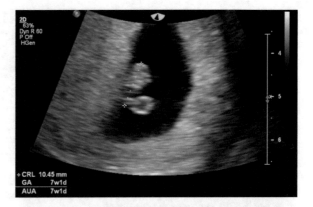

Fig. 1.19 Spectral Doppler with the caliper (gate) positioned on the cardiac structures of a 7-week embryo

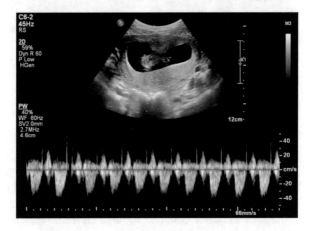

At this early stage, the key investigation to assess the condition of the embryo makes use of ultrasound in order to find the heartbeat and determine the heart rate. PW Doppler can be employed with the caliper (gate), determining the sample volume, positioned on the embryo in order to generate a spectral Doppler trace (Fig. 1.19). This way, any moving structure of the embryonic heart contributes to the signal.

The main ultrasound determinant of the risk of chromosomal abnormalities, such as trisomy 13, 18, and 21, is the nuchal translucency (NT, see Fig. 1.20). This is measured between 11 and 14 weeks of gestation. NT larger than 3.5 mm indicates a higher risk of chromosomal defects, fetal death, and major fetal abnormalities [20]. Values larger than 6 mm may also indicate fetal hydrops, i.e., the accumulation of excessive fluid in some compartments of the fetus [21]. The NT measurement is combined with the CRL assessment, which provides an accurate estimation of the gestational age, and the evaluation of the presence and size of the nasal bone (Fig. 1.20). Altogether, along with the mother's age and the maternal serum concentration of PAPP-A and Beta-HCG, these measurements are used to feed a predictive model for risk assessment [22]. All these early-stage measurements require high

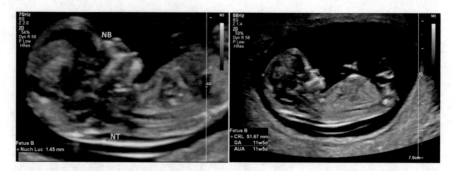

Fig. 1.20 Fetal ultrasound images presenting a measurement (left image) of nuchal translucency (NT) with evaluation of the nasal bone (NB), along with a measurement (right image) of crown-rump length (CRL) to assess the gestational age

resolution; therefore, they are performed by high-frequency (7–9 MHz), transvaginal ultrasound.

1.4.2   Second Trimester Scans

Week 14 marks the start of the second trimester. We are no longer referring to an embryo, but rather to a fetus that is rapidly developing in its anatomy and function.

All organs are developing, bones are getting stronger, and fat is starting to accumulate. At this stage, it is possible to evaluate the fetal development and condition in detail by transabdominal ultrasound. Such evaluation is usually carried out around 19 weeks of gestation throughout the so-called *structural ultrasound scan*. This extensive examination combines several measurements that are detailed in the following.

1.4.2.1   Fetal Biometry

B-mode transabdominal ultrasound with a convex array is usually employed to image and measure the fetal organs. A number of measurements are performed and the results compared to the normal values, indicating good development of the fetus. The main measurements are listed hereafter:

- Biparietal diameter (BPD): this is the maximum distance across the scalp in the lateral direction (Fig. 1.21).
- Head circumference (HC): this is the approximation of the head circumference made by an overlaid ellipse (Fig. 1.21). Additional structures in the fetal head are also evaluated, such as the anterior and posterior horns, the choroid plexuses, the cavum septum pellucididum (CSP), the fetal cerebellum, and the cisterna magna (Fig. 1.21).

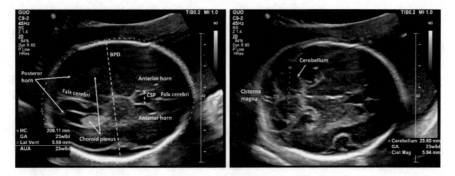

Fig. 1.21 Ultrasound images of the fetal head (orientation: nuchal side on the left and frontal side on the right) across two different planes. On the left image, the measurement of the biparietal diameter (BPD) and the head circumference (HC) are shown, along with the measurement of the lateral ventricle and the cavum septum pellucididum (CSP). Several structures can also be recognized in this view that are reported on the image. On the right image, a different plane is shown that is adopted for the measurement of the cerebellum and the cisterna magna

Fig. 1.22 Ultrasound view for the measurement of the abdominal circumference (AC). In this view, the stomach, the umbilical vein, the aorta (Ao), and the spine are also visible

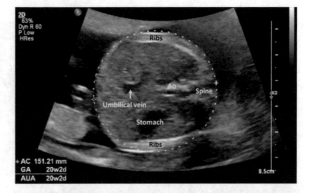

Fig. 1.23 Ultrasound image showing the measurement of the femoral length (FL)

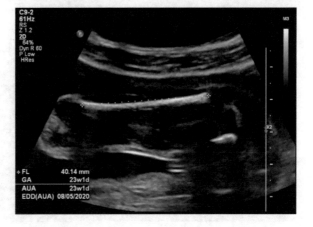

- Abdominal circumference (AC): the abdominal circumference is estimated by overlying an ellipse on the largest abdominal contour (Fig. 1.22). Additional structures visible and evaluated in this plane are the stomach and the umbilical vein (Fig. 1.22).
- Femur length (FL): the length of the femoral bone (Fig. 1.23) is an important determinant of the fetal growth. Typical values at 20 weeks are between 30 and 35 mm. Shorter FL may indicate an elevated risk of abnormalities.

Next to these listed measurements, and the evaluation of the fetal heart presented in the next sections, other structures and organs are also investigated during a structural scan. Figure 1.24 shows two ultrasound images where kidneys, bladder, gallbladder, and diaphragm are visualized and evaluated. Typically, as shown in Fig. 1.25 and Fig. 1.26, spine, face (nose and lips), and limbs are also investigated, along with the bowel. The sex of the fetus can also be determined at this stage of gestation (Fig. 1.26).

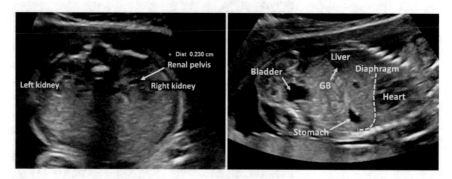

Fig. 1.24 Ultrasound images showing the evaluation of other organs and structures, namely, kidney, bladder, liver, gallbladder (GB), stomach, and diaphragm. The heart is also visible

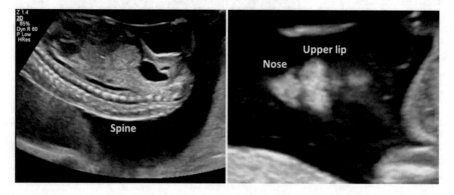

Fig. 1.25 Ultrasound images showing spine, nose, and lips (upper lip) of a fetus

The localization of the placenta is an additional test that is performed during the structural scan. Its position in relation to the cervix is especially relevant in the third trimester. In case the placenta covers the cervix, a cesarean section will be performed.

1.4.2.2   Morphological Evaluation of the Fetal Heart

The circulatory system of a fetus differs from that of an adult in several aspects. Blood is oxygenated through the placenta, which exchanges oxygen and nutrients with the maternal blood. The umbilical vein carries oxygenated blood from the placenta to the right ventricle of the fetus via the inferior vena cava. Along this path, the semifunctional liver is bypassed through the ductus venosus. From the right atrium, oxygenated blood goes into the right ventricle as well as into the left atrium through the foramen ovale (Fig. 1.27). This results into two blood streams, produced by the right and the left ventricle. These streams join back in the aorta, as the pulmonary artery is connected to the aorta via the ductus arteriosus, which shunts blood away from the pulmonary circulation into the aorta. Deoxygenated blood is carried back to the placenta via the umbilical arteries, which are wrapped around the umbilical veins. In conclusion, both ventricles support the systemic circulation, while the transpulmonary circulation is bypassed.

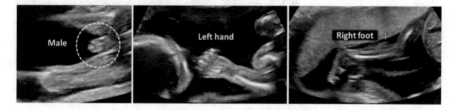

Fig. 1.26 Ultrasound images showing sex and limbs of a fetus

Fig. 1.27 Ultrasound image showing a 4-chamber view of the fetal heart. Several structures are visible: right ventricle (RV), left ventricle, right atrium (RA), left atrium (LA), foramen ovale (FO), intraventricular septum (IVS), moderator band, descending aorta (Ao), the spine, and the pulmonary veins (PVs)

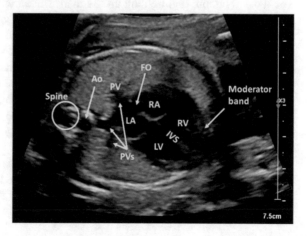

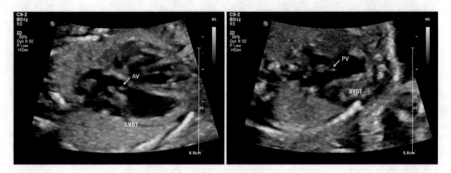

Fig. 1.28 Ultrasound image showing the fetal left ventricle outflow tract (LVOT) and the right ventricle outflow tract (LVOT). The aortic valve (AV) and pulmonary valve (PV) are also visible

Fig. 1.29 Fetal cardiac outflow assessment by the three-vessel view (3VV), showing the superior vena cava (SVC), the aorta (Ao), and the pulmonary artery (PA)

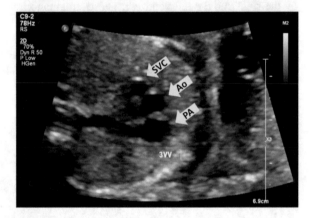

The morphology of the fetal heart can be evaluated in different ultrasound views. The most common, highlighting all the main structures and compartments, is the 4-chamber view (Fig. 1.27). All the 4 cardiac chambers can simultaneously be observed in this view, including the foramen ovale and the descending aorta. The moderator band is often used to guide toward a proper view of the right ventricle. Besides the blood pools, this view also permits assessment of the wall thickness in order to diagnose cardiac hypertrophy. To evaluate the ventriculoarterial connections, the left and right cardiac outflow tracts are visualized (Fig. 1.28). The final plane to assess the cardiac outflow is the so-called three-vessel view (Fig. 1.29). In this view, the size and position of the pulmonary artery the aorta and the superior vena cava are evaluated.

For a better assessment of the vascular structures, the aortic arch is also investigated (Fig. 1.30a). This view presents all three neck and head vessels, which depart from the aortic arch. Heart and spine are also visible. Further investigation involves a functional assessment by Doppler ultrasound and, in particular, color Doppler

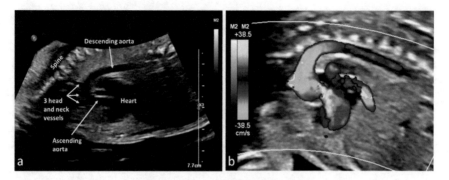

Fig. 1.30 Ultrasound view of the aortic arch in B-mode (a) and color Doppler (b)

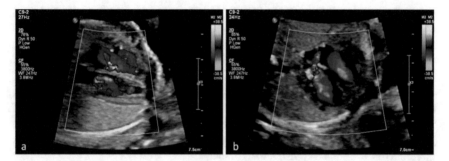

Fig. 1.31 Color Doppler of a fetal heart in a transversal view (a), with focus on interventricular septum, and in a 4-chamber view (b), with visualization of the flow pattern across all the four chambers

analysis. Figure 1.30b shows a corresponding color Doppler image of the aortic arch, but several relevant views are used for functional assessment of the fetal heart (next section).

The morphological analysis of the fetal heart, followed by the functional analysis (next section), are especially intended to diagnose congenital heart disease. This is the most common cause of major congenital anomalies, with a live birth incidence of about 1% [23]. In almost 10% of cases, congenital heart disease is associated with chromosomal abnormalities, such as trisomies [24].

1.4.2.3 Functional Evaluation of the Fetal Heart

The cardiac function is mostly associated with the produced blood flow. Therefore, Doppler ultrasound is the leading technology for functional assessment of the fetal heart. In particular, color Doppler is employed in a transversal plane to assess the interventricular septum for defects (Fig. 1.31a). Visualization of the blood flow in the different chambers is obtained by a 4-chamber view (Fig. 1.31b), similar to that adopted for the morphological investigation. This view is especially used to evaluate possible turbulence and regurgitation from the cardiac valves. The total flow or cardiac output can instead be evaluated through the aortic arch view (Fig. 1.30b). More detailed assessment of the flow generated by the two ventricles can be obtained by separate visualization of the outflow tracts of the right and left ventricles.

Next to flow and perfusion measurements, the cardiac function can directly be assessed in terms of wall motion. This can be investigated by M-mode ultrasound, which is especially suitable to analyze fast-moving structures like the myocardium and the cardiac valves. M-mode ultrasound can also be used for accurate assessment of the fetal heart rhythm.

1.4.2.4 Liquor Volume

Estimation of the liquor volume is a fundamental step in the assessment of fetal condition. The amniotic fluid that bathes the fetus is necessary for its proper development. The average volume increases with gestational age, peaking at 0.4–1.2 L during the third trimester (weeks 34–38 of gestation) [25]. An inadequate volume of amniotic fluid is referred to as oligo- or anhydramnios (depending on the amount of amniotic fluid). This condition, often caused by impaired placental function, can result in poor development of the lungs and the limbs. On the contrary, an excess of amniotic fluid, referred to as poly-hydramniosis, may be associated with possible genetic disorders, fetal abnormalities, such as esophageal or duodenal atresia or maternal disease such as (gestational) diabetes.

Besides a subjective assessment of liquor volume, favored by the hypoechoic nature of the amniotic fluid, some indexes are generally adopted for a quantitative assessment. The maximum deepest pool (MPD) is the longest distance between the fetus or placenta and the uterine wall (Fig. 1.32). The amniotic fluid index is an

Fig. 1.32 Assessment of liquor volume by measurement of the maximum deepest pool (MPD)

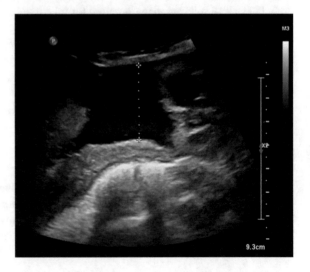

additional measure of liquor volumes that accounts for multiple sides. This is calculated as the sum of the deepest pool in each quadrant of the uterus. These values are used to evaluate the liquor volume as compared to normal values.

1.4.3   Third Trimester and Labor

1.4.3.1   Cardiotocography (CTG)

As shown in Fig. 1.33, cardiotocography (CTG) is the simultaneous and continuous recording of fetal heart rate and uterine contractions [26]. It is performed antenatally to determine the optimal timing and mode of delivery in pregnancies that are considered at risk. It is also performed during labor to monitor the fetal condition [27]. Indeed, it has been shown that the use of the CTG is associated with a decrease in neonatal mortality [28].

Continuous CTG recording of the fetal heart rate makes use of Doppler ultrasound, where the ultrasound probe is positioned on the maternal abdomen and maintained in a fixed position by means of an elastic belt (Fig. 1.33). The main contribution to the extracted Doppler signal originates from moving cardiac structures rather than flow. The resulting Doppler signals often show poor signal-to-noise ratio, making the interpretation of the Doppler recordings very challenging. This is especially severe for premature deliveries, high body-mass-index (BMI) mothers, multiple pregnancies, and during the second stage of labor [26]. As a result, the extracted Doppler signal requires dedicated algorithms for robust estimation and interpretation of the heart rate [29]. This usually requires the definition of a time window where an average heart rate is estimated.

Despite the employment of advanced signal processing, frequent periods of signal loss can still be observed due, e.g., to fetal movements or probe displacement. As a result, the Federation of Gynecology and Obstetrics (FIGO) recommends to only use signals for clinical analysis when no more than 20% of the recording time

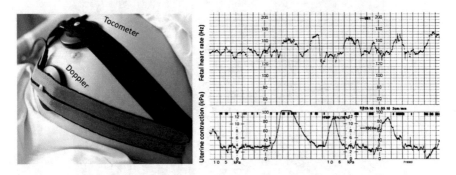

Fig. 1.33 Cardiotocographic registration by means of Doppler ultrasound and tocography to estimate fetal heart rate and the uterine contractions

is lost [30]. This situation demands continuous attention by the operators who need to optimally reposition the probe as soon as signal loss is experienced.

In order to increase the field of view and achieve better coverage of the cardiac structures, the CTG Doppler probe is commonly made of multiple single-element transducers of 1 cm in diameter, which are all connected together. A typical geometry consists of a central element surrounded by multiple elements in a circular fashion [31, 32]. The resulting sample volume is rather large, whereas the pressure field is complex and irregular [26]. Recent research has shown the development of systems for continuous tracking of the fetal heart, with the aim of reducing the need for operator interaction as well as for improving the signal quality while minimizing the acoustic energy transferred to the fetus [33, 34].

1.4.3.2 Umbilical Cord Doppler

Ultrasound Doppler of the umbilical artery (Fig. 1.34) may be used for monitoring the fetal well-being in the late second and third trimester of pregnancy. It is especially indicated in case of fetal growth restriction, along with the Doppler investigation of the middle cerebral artery (Fig. 1.16). Abnormal blood flow through the umbilical artery or the middle cerebral artery in case of fetal growth restriction is a sign of placental insufficiency.

As shown in Fig. 1.34, spectral Doppler is adopted for this investigation, followed by analysis of the resulting waveform in order to extract a number of parameters of diagnostic value. The most common parameters are the ratio between systolic and diastolic velocity, the pulsatility index, calculated as the difference between peak systolic velocity and end diastolic velocity normalized with respect to the time-averaged velocity, and the resistive index, calculated as the difference

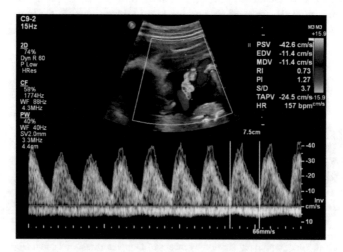

Fig. 1.34 Color Doppler map and spectral Doppler (below) of the umbilical cord

between peak systolic velocity and end diastolic velocity normalized with respect to the peak systolic velocity. Color or power Doppler can be used to guide for correct positioning of the caliper (gate) to perform spectral Doppler analysis. In case of abnormal umbilical artery Doppler, further investigation is indicated in order to assess the degree of placental insufficiency.

1.4.3.3   Fetal Orientation

Toward the end of pregnancy, most fetuses move into a position that favors delivery. Normally, the presentation of a fetus is head first (cephalic presentation), facing backward (occiput anterior) with face and body angled to one side and the neck flexed. Abnormal presentations include breech and shoulder, depending on the anatomical part facing the birth canal through the pelvic bones. The fetal position may also be abnormal, e.g., when the fetus is facing forward (occiput posterior).

In preparation for delivery, the exact fetal presentation and position must be determined, such that proper decision making can be made in relation to the procedure adopted for delivery. Therefore, transabdominal ultrasound imaging can be performed in order to determine the fetal position relative to the spine and the pelvic bones. Dedicated image analysis tools have also been developed that can assist with the interpretation of the ultrasound images and assessment of the fetal descent through the birth canal [35].

1.4.3.4   Cervical Length

The cervix undergoes a progressive shortening toward labor, ending with its opening during labor, which favors the passage of the baby's head through the birth canal into the vagina. Therefore, the measurement of the cervical length is a fundamental test in order to assess the risk of preterm delivery. A cervical length that is shorter than 30 mm before 34 weeks of gestation is considered to be associated with an

Fig. 1.35 Measurement of the cervical length by transvaginal ultrasound

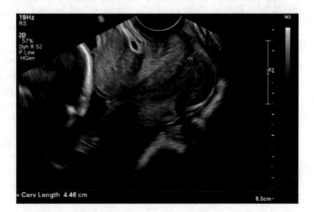

increased risk of preterm labor [36]. As shown in Fig. 1.35, assessment of the cervical length is usually performed by (high-frequency) transvaginal ultrasound imaging, as the probe can directly be positioned on the side of the cervix, producing high-quality and high-resolution images.

References

1. Webster, J.G.: Medical instrumentation-application and design. J. Clin. Eng. **3**(3), 306 (1978)
2. Álvarez-Arenas, T.E.G.: Acoustic impedance matching of piezoelectric transducers to the air. IEEE Trans. Ultrason. Ferroelectr. Freq. Control. **51**(5), 624–633 (2004)
3. Cobbold, R.S.C.: Foundations of Biomedical Ultrasound. Oxford University Press, New York (2007)
4. Mischi, M., Rognin, N., Averkiou, M.: Ultrasound imaging modalities. In: Comprehensive Biomedical Physics, pp. 323–341. Oxford, Elsevier (2014)
5. Thijssen, J.M., Mischi, M.: Ultrasound imaging arrays. In: Comprehensive Biomedical Physics, pp. 361–386. Oxford, Elsevier (2014)
6. Harris, F.J.: On the use of windows for harmonic analysis with the discrete Fourier transform. Proc. IEEE. **66**(1), 51–83 (1978)
7. Tanter, M., Fink, M.: Ultrafast imaging in biomedical ultrasound. IEEE Trans. Ultrason. Ferroelectr. Freq. Control. **61**(1), 102–119 (2014)
8. Montaldo, G., Tanter, M., Bercoff, J., Benech, N., Fink, M.: Coherent plane-wave compounding for very high frame rate ultrasonography and transient elastography. IEEE Trans. Ultrason. Ferroelectr. Freq. Control. **56**(3), 489–506 (2009)
9. Jensen, J.A., Nikolov, S.I., Gammelmark, K.L., Pedersen, M.H.: Synthetic aperture ultrasound imaging. Ultrasonics. **44**, e5–e15 (2006)
10. Nyborg, W.L.: Biological effects of ultrasound: development of safety guidelines. Part II: general review. Ultrasound Med. Biol. **27**(3), 301–333 (2001)
11. Rognin, N.G., Frinking, P., Costa, M., Arditi, M.: In-vivo perfusion quantification by contrast ultrasound: Validation of the use of linearized video data vs. raw RF data. In: *Proceedings - IEEE Ultrasonics Symposium*, Beijing, China, pp. 1690–1693 (2008)
12. Wagner, R.F., Insana, M.F., Wagner, D.G.: Statistical properties of radio-frequency and envelope-detected signals with applications to medical ultrasound. J. Opt. Soc. Am. A. **4**(5), 910–922 (1987)
13. Bohs, L.N., Trahey, G.E.: A novel method for angle independent ultrasonic imaging of blood flow and tissue motion. I.E.E.E. Trans. Biomed. Eng. **38**(3), 280–286 (1991)
14. DeVore, G.R., Falkensammer, P., Sklansky, M.S., Platt, L.D.: Spatio-temporal image correlation (STIC): new technology for evaluation of the fetal heart. Ultrasound Obstet. Gynecol. **22**(4), 380–387 (2003)
15. Bonnefous, O.: Time domain formulation of pulse-Doppler ultrasound and blood velocity estimation by cross correlation. Ultrason. Imaging. **8**(2), 73–85 (1986)
16. Kasai, C., Namekawa, K., Koyano, A., Omoto, R.: Real-time two-dimensional blood flow imaging using an autocorrelation technique. IEEE Trans. Sonics Ultrason. **32**(3), 458–464 (1985)
17. Loupas, T., Gill, R.W., Powers, J.T.: An axial velocity estimator for ultrasound blood flow imaging, based on a full evaluation of the Doppler equation by means of a two-dimensional autocorrelation approach. IEEE Trans. Ultrason. Ferroelectr. Freq. Control. **42**(4), 672–688 (1995)
18. Rubin, J.M., Bude, R.O., Carson, P.L., Bree, R.L., Adler, R.S.: Power Doppler US: a potentially useful alternative to mean frequency-based color Doppler US. Radiology. **190**(3), 853–856 (1994)
19. Robinson, H.P., Fleming, J.E.E.: A critical evaluation of sonar 'crown-rump length' measurements. BJOG Int. J. Obstet. Gynaecol. **82**(9), 702–710 (1975)

20. Wright, D., Kagan, K.O., Molina, F.S., Gazzoni, A., Nicolaides, K.H.: A mixture model of nuchal translucency thickness in screening for chromosomal defects. Ultrasound Obstet. Gynecol. **31**(4), 376–383 (2008)
21. Souka, A.P., Krampl, E., Bakalis, S., Heath, V., Nicolaides, K.H.: Outcome of pregnancy in chromosomally normal fetuses with increased nuchal translucency in the first trimester. Ultrasound Obstet. Gynecol. **18**(1), 9–17 (2001)
22. Kagan, K.O., Wright, D., Valencia, C., Maiz, N., Nicolaides, K.H.: Screening for trisomies 21, 18 and 13 by maternal age, fetal nuchal translucency, fetal heart rate, free -hCG and pregnancy-associated plasma protein-A. Hum. Reprod. **23**(9), 1968–1975 (2008)
23. Van Der Linde, D., et al.: Birth prevalence of congenital heart disease worldwide: a systematic review and meta-analysis. J. Am. Coll. Cardiol. **58**(21), 2241–2247 (2011)
24. van Velzen, C., et al.: Prenatal detection of congenital heart disease-results of a national screening programme. BJOG Int. J. Obstet. Gynaecol. **123**(3), 400–407 (2016)
25. Fischer, R.L.: Amniotic Fluid: Physiology and Assessment. Glob. Libr. Women's Med. (2008) https://doi.org/10.3843/GLOWM.10208
26. Hamelmann, P., et al.: Doppler ultrasound technology for fetal heart rate monitoring: a review. IEEE Trans. Ultrason. Ferroelectr. Freq. Control. **67**(2), 226–238 (2020)
27. Ayres-de-Campos, D., Spong, C.Y., Chandraharan, E.: FIGO consensus guidelines on intrapartum fetal monitoring: Cardiotocography. Int. J. Gynecol. Obstet. **131**(1), 13–24 (2015)
28. Ananth, C.V., Chauhan, S.P., Chen, H.-Y., D'Alton, M.E., Vintzileos, A.M.: Electronic fetal monitoring in the United States. Obstet. Gynecol. **121**(5), 927–933 (2013)
29. Signorini, M.G., Fanelli, A., Magenes, G.: Monitoring fetal heart rate during pregnancy: contributions from advanced signal processing and wearable technology. Comput. Math. Methods Med. **2014**, 707581 (2014)
30. Ayres-de-Campos, D., Bernardes, J.: Twenty-five years after the FIGO guidelines for the use of fetal monitoring: time for a simplified approach? Int. J. Gynecol. Obstet. **110**(1), 1–6 (2010)
31. Hamelmann, P., et al.: Improved ultrasound transducer positioning by fetal heart location estimation during Doppler based heart rate measurements. Physiol. Meas. **38**(10), 1821–1836 (2017)
32. Kribèche, A., Tranquart, F., Kouame, D., Pourcelot, L.: The Actifetus system: a multidoppler sensor system for monitoring fetal movements. Ultrasound Med. Biol. **33**(3), 430–438 (2007)
33. Hamelmann, P., Mischi, M., Kolen, A., van Laar, J., Vullings, R., Bergmans, J.: Fetal heart rate monitoring implemented by dynamic adaptation of transmission power of a flexible ultrasound transducer array. Sensors. **19**(5), 1195 (2019)
34. Hamelmann, P., Vullings, R., Mischi, M., Kolen, A.F., Schmitt, L., Bergmans, J.W.M.: An extended Kalman filter for fetal heart location estimation during Doppler-based heart rate monitoring. IEEE Trans. Instrum. Meas. **68**(9), 3221–3231 (2019)
35. Casciaro, S., et al.: Automatic evaluation of progression angle and fetal head station through intrapartum echographic monitoring. Comput. Math. Methods Med. **2013**, 278978 (2013)
36. Van Baaren, G.J., et al.: Predictive value of cervical length measurement and fibronectin testing in threatened preterm labor. Obstet. Gynecol. **123**(6), 1185–1192 (2014)

Chapter 2
CRS4 Telemed: Open and Low-Cost Technologies for Real-Time Telesonography

Francesca Frexia, Vittorio Meloni, Mauro Del Rio, and Gianluigi Zanetti

Contents

2.1   Introduction

Telemedicine is generally considered to be a promising approach to improve the quality and uniformity of clinical care, at national and international levels [1–7]. The application of telemedicine is also seen as a valid strategy to decrease the costs of medical services, for example, supporting organisational paradigms such as the hub-and-spoke model [8–10], where services are delivered into networks consisting of a main specialised centre (hub), complemented by several secondary territorial centres (spokes). The origins of telemedicine can be traced back to the first experiments in transmitting biomedical signals and data via telegraph, followed by the use in situations where remote assistance was the only way to obtain medical support, such as long sea crossings or space travels [11]. At present, a wide range of different

F. Frexia (✉) · V. Meloni · M. Del Rio · G. Zanetti (Deceased)
CRS4 - Center for Advanced Studies, Research and Development in Sardinia,
Pula (CA), Italy
e-mail: francesca.frexia@crs4.it; vittorio.meloni@crs4.it; mauro.delrio@crs4.it

© Springer Nature Switzerland AG 2021
D. Pani et al. (eds.), *Innovative Technologies and Signal Processing in Perinatal Medicine*, https://doi.org/10.1007/978-3-030-54403-4_2

applications falls under the telemedicine "umbrella", for example, home monitoring, remote surgical operations or second opinion consultations. This variety is reflected in the difficulty in providing a single definition of telemedicine, resulting in broad sets of definitions coined over the years. In 1971, the first formal and published definition characterised telemedicine as *"the practice of medicine without the usual physician-patient confrontation...via an interactive audio-video communications system"* [12]. The World Health Organization has adopted in 1998 the following broad description [13]: *"The delivery of health care services, where distance is a critical factor, by all healthcare professionals using information and communication technologies for the exchange of valid information for diagnosis, treatment and prevention of disease and injuries, research and evaluation, and for the continuing education of health care providers, all in the interests of advancing the health of individuals and their communities"*. Afterwards, many attempts have followed to delineate more precisely scopes and boundaries of telemedicine, up to arriving to 104 different peer-reviewed definitions compared in a study in 2007 [14], mapping different contexts and perspectives. The difference between telemedicine and tele-health was also analysed, indicating the former as focusing mainly in the clinical aspects, such as the delivering of clinical care and medical professional education, while the latter as a wider area covering a greater number of application, not necessarily clinical, such as the sharing of administrative information or social aid [15, 16].

From a technological point of view, three main categories can be considered, according to the types and configurations of transmission [17, 18]:

- *Store and forward*: also called asynchronous, it consists in the acquisition of the data during an examination followed by their further sharing with healthcare professionals.
- *Real-time*: also called synchronous, it enables the live interaction between the operators during the clinical evaluation.
- *Remote monitoring*: where sensors and devices are used to track and control health and vital signs of a person, in a synchronous or asynchronous way.

The technological choices must follow the needs expressed by the clinical use cases. For example, in situations of second opinion related to standard diagnostic techniques (e.g. laboratory tests, radiological images), store and forward should be preferred, while remote monitoring is widely used for patients affected by chronic diseases. On the other hand, real-time telemedicine is required when the communication between the remote participants has to be continuous. Typical cases are all the operator-dependent diagnostic techniques, which require a specific competence of the operator who performs the examination as a necessary condition to obtain a correct diagnosis. Real-time transmission, though, can be costly and generally needs dedicated networks and/or commercial solutions, which are often proprietary, closed and expensive [19].

In this chapter, we will illustrate *CRS4 Telemed*, the telemedicine application we developed, and its evolution in over 10 years. The system is based on low-cost hardware and open-source software and enables a specialist to guide a remote examination in real-time, supporting the direct interaction between the clinicians involved.

2.2 CRS4 Real-Time Telemedicine System

CRS4 Telemed is a system that allows two clinicians, a sonographer and a specialist, to collaborate remotely in performing a real-time ultrasound examination on a patient. The sonographer can be an operator with no experience in the specific field or a specialist who needs to discuss a particular case with a remote colleague. The sonographer operates the ultrasound device, while the specialist is geographically far from the patient. The system enables the communication between the doctors via a VoIP system and allows the specialist to watch, using a pc or a tablet, the audio/video streams of the ultrasound machine and of a webcam that films the scene of the examination. In this way, the specialist can instruct the sonographer on how to perform the examination as if they were in the same room with the patient. CRS4 Telemed has been developed using COTS (Commercial Off the Shelf) technologies, which can facilitate the adoption of the system in centres with limited economic resources. The steps to require and perform a teleconsultation are depicted in Fig. 2.1. A series of videos on CRS4 YouTube channel illustrates the functioning of the system.[1,2]

2.2.1 Evolution

Three main milestones can be identified in the development of the system:

- *First prototype (2009)*: it was a proof-of-concept that demonstrated the feasibility of supporting effectively the audio/video communication with a prototype obtained assembling hardware not specifically designed for that kind of applications. In this version, both the specialist and the sonographer were equipped with a notebook. These initial tests revealed a bandwidth requirement of 700 kb/s, proving that the prototype and the network infrastructure at that time could support an audio/video streaming with quality sufficient to perform a diagnosis [20].
- *Complete desktop platform (2010–2014)*: we evolved the prototype developing a more sophisticated version composed of two softwares: a Mac OS X desktop application, used by the specialist to watch the streams and talk to the sonographer, and an iOS app for the sonographer to communicate with the specialist. In addition, a central server is responsible for handling the communication between the clinicians and for the configuration of the system. This version was clinically tested in paediatric cardiology [21] and in emergency department [22].

---

[1] https://www.youtube.com/watch?v=9ynwy28kKnQ

[2] https://www.youtube.com/watch?v=xSfEtZKyyHc

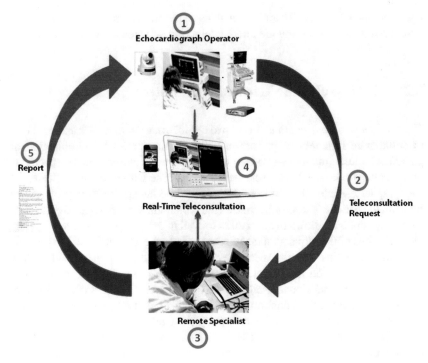

Fig. 2.1 The clinical process supported by the system includes all the steps to require a teleconsultation, perform it and compile a structured report about the examination

- *Mobile connectivity and augmented reality (2014–2018)*: in this phase we moved the implementation to mobile technologies with the development of an Android application for tablets and smartphones. This version of the system was successfully tested in conditions of total mobility using the Long Term Evolution (LTE) networks. We also explored the use of augmented reality [23] to test if the state-of-the-art technologies in this field could improve the communication between the clinicians. Our use case considered the effect in the interaction when the sonographers were equipped with smartglasses, enabling them to see a pointer guided by the specialist to indicate some regions of interest.

2.2.2 Video Acquisition and Streaming

As mentioned before, the specialist guides the sonographer by watching two streams: the output of the ultrasound machine and the video of a network camera that records the scene, as shown in Fig. 2.2.

The first stream allows the specialists to see the ultrasound machine output as if they were performing the examination, while the camera video helps the specialist to control several external factor, such as how the operator is moving the probe, the

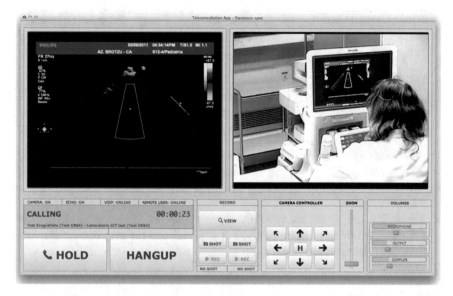

Fig. 2.2 Desktop interface of the specialist application

exact position of the patient and their external look. These streams are generated by COTS hardware: a video encoder and a network camera. The first one is a device that acquires, optionally digitises (in case of devices with analog output), encodes and transmits the output from the ultrasound machine. The acquisition depends on the output supported by the ultrasound (VGA, composite, HDMI, etc.): older machines usually use composite output, while modern ones support HDMI. The network camera is a simple video camera, like those used for video surveillance, that streams the scene of the exam. Both types of devices can support, depending on the model chosen, different encoding formats (M-JPEG, MPEG, H.264, H.265), resolutions (640 × 480, 720p, 1080p, etc.) and streaming protocols (RTSP, HTTP). For our purpose, the images, especially those coming from the encoder, must guarantee a quality sufficient to the specialist to make a diagnosis.

2.2.3   Architecture

CRS4 Telemed has been designed to support the hub-and-spoke collaboration model implemented via a variety of network configurations, such as wide geographical areas between different clinical centres connected via the Internet or different departments within the boundaries of the same hospital's LAN. The system is logically divided in three main subsystems, as depicted in Fig. 2.3:

- *Ultrasound station*: it is the subsystem that asks a remote teleconsultation and transmits the audio/video streams.

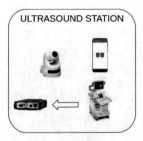 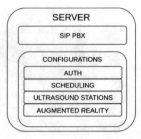 

Fig. 2.3 Components of CRS4 Telemed

- *Specialist station*: it is the component supporting the specialist to perform the teleconsultation.
- *Server*: it is the core of the system, responsible for the configurations (e.g. users, patients, streaming devices, ultrasound stations) and for the VoIP communications establishment and handling.

2.2.3.1 Ultrasound Station

The ultrasound station is associated to the room where the examination is held. It has both hardware and software components, which are as follows:

- *Network audio/video encoder*: it is the device connected to the ultrasound machine that exposes the audio/video stream in the network.
- *Network camera*: it is the camera that films the scene. Depending on the model used, it is possible to control the framing and the zoom remotely.
- *Sonographer mobile application*: it is the application used by the sonographer to request and perform a teleconsultation. During the teleconsultation, the sonographer uses the smartphone to talk via VoIP to the specialist and to see the output of the camera, in order to be aware of what the specialist is viewing. It was first implemented as an iOS app and then developed for Android.

The system can support several ultrasound stations, enabling to require teleconsultation from different examination rooms.

2.2.3.2 Specialist Station

The specialist station is the hw/sw component, enabling the specialist to respond to requests for consultations and to see the streams of the examination. The software application presents a list of pending teleconsultation requests, which can be selected by the specialist to initiate the audio and video communication with the sonographer. The first version of the software was a Mac OS X Desktop application, which also allowed the specialist to move or zoom the framing of the camera, if the ultrasound station supported the feature.

Fig. 2.4 Specialist's
mobile app

The second version has been implemented for Android tablets (Fig. 2.4), also adding new features, such as video resizing and network camera stream deactivation. This feature is particularly important to reduce the required bandwidth in case of poor connection or to hide the patient identity when the system is used for educational purposes. In the last version of the system, the specialist station also includes the possibility of indicating regions of interest to be displayed to the operator as augmented reality indications.

2.2.3.3 Server

The server has two main subcomponents. The first is responsible for the audio communication, as it is an IP PBX and VoIP gateway, that is a system which establishes, manages and terminates audio communications using the Session Initiation Protocol (SIP) [24]. The second component is the Application Server, managing the main configurations of the system, which are:

- *User Authentication*: the sonographers and the specialists are registered into the system in order to login into the apps. This section configures the accounts of the user. The protocol used for authentication is oAuth [25].
- *Stations Management*: for every Ultrasound Station, this section contains the parameters of the video encoder and the network camera, such as the IP address, the URL of the stream and the authentication configuration to access the stream.
- *Scheduling*: this section contains data related to teleconsultations, such as the sonographer and the specialist involved, the transmitting ultrasound station and the status of the teleconsultation (e.g. requested, established and terminated).
- *SIP*: this section contains the configuration of the SIP server and users. Each user of the system has an associated account for the IP PBX server, which is necessary to route the voice communication correctly.
- *Augmented Reality*: this section configures parameters for the augmented reality.

2.2.3.4   Workflow

The basic flow to establish and perform a teleconsultation is shown in Fig. 2.1.

1. The sonographer logs into the app on the smartphone.
2. The sonographer creates a teleconsultation request by selecting the Ultrasound Station from the list provided by the Application Server and entering the patient's demographic data. After that, the waiting for the specialist's answer starts.
3. The specialist logs into the app in the tablet, which displays the list of teleconsultation requests. By selecting one of them, the app shows the teleconsultation page where the user can initiate the VoIP call. When the sonographer answers the call in the app the examination can proceed.
4. The specialist's app displays the video streams and the teleconsultation is held by the clinicians until one of them finishes the VoIP communication.
5. After the examination, the specialist can compile a report with the outcomes.

2.2.3.5   Augmented Reality

The clinical trials, better described in Sect. 2.3, showed that the system is able to support an efficient interaction during teleconsultation, but also highlighted that communication occasionally can suffer from a lack of visual assistance, especially with sonographers not experienced in the specific clinical domain. In some cases, the specialists try to indicate a particular region of the screen, even knowing that the sonographer can't see their movements. These experimental observations motivated us to explore the usage of Augmented Reality (AR) to improve communication between the clinicians. AR adds virtual objects to the real world by overlaying them in real-time to the view of smartphones, PC or smartglasses. The position of the super-imposed object can be obtained by considering different factors, such as geo-location indications [26, 27], sensors on the device, such as accelerometer, magnetometer and gyroscope [27] or by using markers [28]. The markers are distinctive images that are added to the scene, and can be recognised by the device's camera and used to calculate the position of the virtual object in the space.

We applied a marker-based augmented reality solution to our system, introducing two markers on the sides of the screen of the ultrasound and one near its keyboard, as shown in Fig. 2.5. In the AR solution, the sonographer uses smartglasses[3] (which makes the virtual objects appear on the lenses) instead of the smartphone. The markers on the sides of the screen have been added to enable the specialist to attract the operator's attention on precise regions of the ultrasound images. In fact, when the specialist selects a point in the ultrasound machine output displayed on

---

[3] The model used is the Android based Epson Moverio BT-200

Fig. 2.5 An examination performed using the AR solution. Notice the markers on the side of the display and near the keyboard

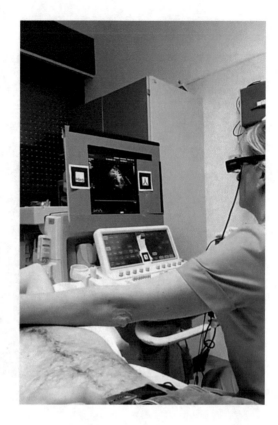

the tablet, an indicator appears on the smartglasses of the remote operator, superimposed to the indicated region. The marker associated to the keyboard has been added as, during the initial tests, we noticed that sometimes inexperienced sonographers could not find the required buttons in the ultrasound keyboard. We have introduced, then, an additional marker for the keyboard, adding to the specialist's app the ability to select a button. When the specialist indicates a button in the tablet app, represented in Fig. 2.6, a pointer appears in the sonographer's smartglasses at the position of this button on the keyboard. The complete AR setup is represented in Fig. 2.7.

The tests showed that the solution was considered as valuable by users, but some critical points were also highlighted. First of all, some additional training is necessary to use smartglasses properly, because the positioning of the operator with respect to the markers is not intuitive. Secondly, for each user, a specific calibration of the device is necessary to overcome some alignment problems between the user's eyes, the virtual objects and the real world. Finally, great attention has to be paid to the ergonomic factors during the device selection, as the smartglasses were considered uncomfortable, especially if used for a long time.

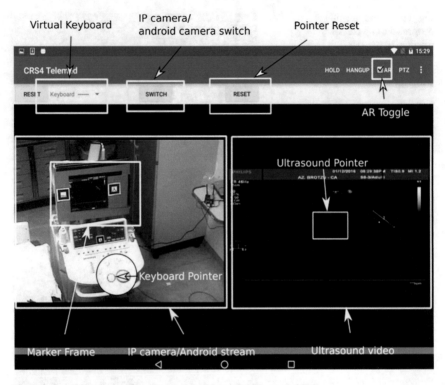

Fig. 2.6 Specialist tablet view. (Reproduced from Del Rio et al. 2018)

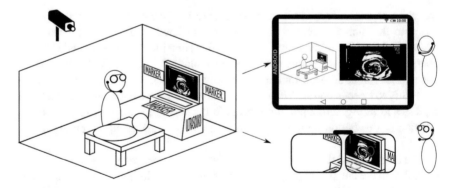

Fig. 2.7 Augmented reality setup

2.3 Evaluating the System: Clinical and Economical Perspectives

In this section, we describe the use cases in which the system has been applied and successfully tested. A preliminary cost-benefit analysis for the use of the system in a regional context will be also presented.

2.3.1 Paediatric Cardiology

Paediatric Cardiology is the use case which inspired the creation of the system, thanks the collaboration with Dr Roberto Tumbarello, Dr Sabrina Montis and the staff of the Paediatric Cardiology Unit at Azienda Ospedaliera "G. Brotzu" in Cagliari (Italy) [20, 21]. Paediatric Cardiology is a perfect example of an operator-dependent diagnostic technique, as in this field, without a specific expertise, even an expert cardiologist could have difficulties in performing effectively the sonographic examination. In 2013, we conducted a clinical trial in the Italian region of Sardinia, to test the system in the support of specialised consultation for Congenital Heart Diseases (CHD), the most common congenital disorders, affecting 6‰ to 13‰ [29–32] live-born children. In face of a population of about 1,600,000 citizens in a territory of 24,090 km^2, with a high presence of CHDs, in Sardinia, there is only one specialised centre in Cagliari. Many children from other parts of the island with suspected CHD need to be transferred to the main centre to perform a specialised examination, increasing the required intervention times to start a treatment. The clinical trial protocol considered 42 cases of children between 6 and 18 years, two expert paediatric cardiologists and an operator acting as sonographer. Each patient was examined twice: the first time by one of the specialists who remotely guided an operator with the telemedicine system, the second time directly by the other specialist. The two clinicians performed the same number of ultrasound examinations, alternating traditional and teleconsultation approach and guiding the same sonographer in the second case. The study showed a 97.6% agreement rate on diagnosis obtained using the teleconsultation system or directly examining the patient.

Total mobility tests were also performed in an educational context, during the *2nd Course of Fetal Echocardiography and Perinatal Cardiology* (2016) in Cagliari: in this case, the system was used to connect the conference venue with the ultrasound station at Brotzu Hospital. Both clinicians were specialists in Paediatric Cardiology and used the system to perform a collaborative diagnosis on very difficult cases. The screen of the specialist's tablet was projected in the conference room to enable the students to follow each step of the examination. In order to respect the privacy of the patients, the network camera was obscured. The tests showed that the system is able to support effectively a remote consultation also using LTE networks.

2.3.2 Point-of-Care Ultrasonography

A second study [22] was guided by Dr Floriana Zennaro and involved the Radiology and the Emergency Departments of the Institute for Maternal and Child Health IRCCS Burlo Garofolo. The study aimed to evaluate the diagnostic accuracy of the Point-of-care Ultrasonography (POC US) examination of children accessing the emergency department, when performed by paediatricians under the remote guidance of expert radiologists.

The study was conducted on a cohort of 52 children aged from 0 to 18 years old, with a total of 170 examinations, and involved 4 paediatric consultants and 5 paediatric radiologists. The protocol envisaged three ultrasound examinations for each case. The first was carried out by a paediatrician in the emergency room telementored by a senior paediatric radiologist using the telemedicine system (TELE POC); the second was performed by the same senior radiologist (UNBLIND RAD) and the third by a second senior paediatric radiologist who didn't know the results of the first two examinations (BLIND RAD).

The research showed one false negative when comparing TELE POC vs UNBLIND RAD and two false negatives contrasting TELE POC with BLIND RAD. The inter-rater agreement [33] among the 9 clinicians that participated in the study reached an excellent value of 0.97. The quality of the transmission was never rated as poor by the users.

2.3.3   Preliminary Cost-Benefit Analysis

The system was also evaluated from an economical point of view, with a preliminary cost-benefit analysis of the potential savings in transfer costs, supposing the system was applied to the Paediatric Cardiology use case at a regional level, for 1 year, in Sardinia [34]. The study simulated the creation of a regional hub-and-spoke network, with Brotzu Hospital acting as the hub providing specialised consultancy, via the telemedicine system, to 8 centres in the main hospitals of the island. The study considered two situations: emergencies, when the patient is transferred from a secondary health facility to the main centre by ambulance, and routine visits, when the patient reaches the specialised centre by private car. The analysis took into account the real data for the consultations on CHD required by other structures to Brotzu in 2012. To calculate the costs without the telemedicine system, we took into account the costs of the vehicle (fuel + fixed costs) and, in case of emergencies, the cost of the ambulance team. We also evaluated the amount of money that could have been saved by avoiding unnecessary transportation, that is, the ones where the patient was negative for CHD. The study estimated potential savings of about 66% of the transfer costs, showing that the effects are significant even considering only a very specific component of the potential savings.

2.4   Discussion

The initial motivations for the prototype development included the need of a solution with limited costs, working in absence of dedicated network and based on technologies, easy to be integrated with the hospital technical equipment. When we started the feasibility study, a typical teleconference system enabling a real-time interaction had a cost about ten times higher than our system [34, 35]. In the years, the offer of telemedicine services increased and improved; therefore, at present, a series of solutions could be analysed and compared to our system to evaluate the most suitable to the requirements on the specific use case [36–38].

During its evolution, the system has been updated and adapted to exploit the potential offered by the technological advancement and to improve its usability, according to the positive and negative feedback from the clinical personnel. For example, the first prototype suffered from ergonomics issues in the sonographer station, as the used notebook was difficult to place close to the ultrasound machine. Also the audio communication presented some issues, since the sonographer had to use the internal microphone of the notebook and this sometimes caused interference with the audio of the ultrasound and the audio returns from the specialist side. These problems were solved with the adoption of smartphone and headphones, which did not affect the mobility of the sonographer during the exam. Some doubts were also related to the usage of tablets for diagnostic purpose, but, at present, several studies have reported that the use of tablet displays does not decrease the accuracy of radiological diagnosis [39, 40], even if there is still a gap in the legislation to be covered regarding the adoption of mobile devices in radiology.

The issues associated to the usage of the system are not only technological: for example, from a legal point of view, using the system in a real clinical environment, all the chain of responsibilities should be clarified to trace the boundaries between the responsibility of the operator performing the examination and the remote specialist [41].

The results obtained by this prototype demonstrated its validity as an economic solution to improve the quality of care quality and the coverage of specialised diagnostic service across a regional territory. The next natural steps to translate the results of our research to clinical practice are the certification and the industrialisation of the prototype, which require the presence of an industrial partner. This could be, for example, an additional service that companies producing ultrasound machines could offer to their customers. As the whole platform is available open-source, also a public institution, like the government of a Region, or a network of health institutions could start from the results obtained, certifying the solution and using the platform. From the research point of view, further trials of the platform wouldn't improve significantly the evidence of the validity of the system, and, therefore, our next plans don't include further developments or validation studies without a partner interested in covering the last mile to move the system to a real clinical environment.

2.5 Conclusion

In this chapter, we presented CRS4 Telemed, a real-time telemedicine system to perform collaborative ultrasound examinations at a distance. The system is based on open-source software and low-cost technology, it was clinically validated in two specific disciplines—paediatric cardiology and emergency radiology—and can be used in all the diagnostic procedures which require direct consultation during the examination of a video stream. The system can also be used for educational purposes, avoiding the presence of many students in an examination room during a clinical evaluation. After a 10-year evolution, the research activities connected to the system are finished and, at present, the platform is ready for an industrialisation process, which could be carried on by private or public subjects, with our direct support. The system is available open-source at https://github.com/crs4/crs4-telemed

Acknowledgements We dedicate this work to the memory of Gianluigi Zanetti, whose legacy continues to inspire. We also want to thank all the persons who supported this work with passion during these years. In particular, Roberto Tumbarello, Sabrina Montis, Andrea Marini, Paola Neroni and all the "G. Brotzu" Hospital Staff in Cagliari; Floriana Zennaro and all the "Burlo Garofalo" Hospital Staff in Trieste; Emiliano Deplano, Carlo Balloi and the "Nostra Signora della Mercede" Hospital Staff in Lanusei; Francesco Cabras, Stefano Leone Monni, Riccardo Triunfo, Cecilia Mascia, Alessandro Piroddi, Luca Lianas, Emanuela Falqui. This work has received funding from Sardinian Regional Authorities under projects ABLE, DIFRA, REMOTE (L.R. 7/2007 year 2009), CONNECT (L.R. 7/2007 year 2012), and from Sardegna Ricerche under project MOST ("Azioni cluster top-down" - POR Sardegna FESR 2007/2013).

References

1. Communication from the Commission to the European Parliament, the Council, the European Economic and Social Committee and the Committee of the Regions on telemedicine for the benefit of patients, healthcare systems and society/COM/2008/0689 final/. (2008) Available via http://eur-lex.europa.eu/legal-content/EN/ALL/?uri=CELEX:52008DC0689
2. European Commission: Market study on telemedicine. Available via https://www.ehealth-news.eu/download/publications/5729-final-report-provision-of-a-market-study-on-telemedi-cine (2018)
3. TELEMEDICINA: Linee di indirizzo nazionali. Available via http://www.salute.gov.it/imgs/C_17_pubblicazioni_2129_allegato.pdf (2014)
4. Alvandi, M.: Telemedicine and its role in revolutionizing healthcare delivery. Am. J. Accountable Care. **5**(1), e1–e5 (2017)
5. Telehealth: A Path to Virtual Integrated Care. Available via https://www.aha.org/system/files/media/file/2019/02/MarketInsights_TeleHealthReport.pdf (2019)
6. Zanaboni, P., Wootton, R.: Adoption of routine telemedicine in Norwegian hospitals: progress over 5 years. BMC Health Serv Res. **16**(1), (2016)
7. New Zeland Telehealth Forum and Resource Centre. Available via https://www.telehealth.org.nz/ (2019)
8. Elrod, J.K., Fortenberry Jr., J.L.: The hub-and-spoke organization design: an avenue for serving patients well. BMC Health Serv Res. **17**(1), 457 (2017)
9. Huddleston, P., Zimmermann, M.: Stroke care using a hub and spoke model with telemedicine. Crit. Care Nurs. Clin. North Am. **26**(4), 469–475 (2014)
10. Switzer, J.A., Demaerschalk, B.M., et al.: Cost-effectiveness of hub-and-spoke telestroke networks for the management of acute ischemic stroke from the hospitals' perspectives. Circ. Cardiovasc. Qual. Outcomes. **6**, 18–26 (2013)
11. Vladzymyrskyy, A., Jordanova, M., Lievens, F.: A century of telemedicine: Curatio Sine Distantia et Tempora. (2016)
12. Bird, K.T.: Teleconsultation: a new health information exchange system. Third Annual Rep. Veterans Admin., Washington, D.C. (1971)

13. WHO Group Consultation on Health Telematics: A health telematics policy in support of WHO's Health-for-all strategy for global health development: report of the WHO Group Consultation on Health Telematics. World Health Organization. Available via DIALOG. https://apps.who.int/iris/handle/10665/63857 (1998)
14. Sood, S., et al.: What is telemedicine? A collection of 104 peer-reviewed perspectives and theoretical underpinnings. Telemed. e-Health. **13**(5), 573–590 (2007)
15. Maheu, M.M., Whitten, P., Allen, A.: E-Health, Telehealth, and Telemedicine: a Guide to Start-up and Success, 1st edn. Jossey-Bass Inc., San Francisco (2001)
16. American Nurses' Association: Developing Telehealth Protocols: A Blueprint for Success. American Nurses Association, Washington, D.C. (2001)
17. World Health Organization: Telemedicine – Opportunities and developments in Member States. World Health Organization (2010)
18. Telehealth and Telemedicine. (2018) Available via http://www.ehealthwork.eu/FC/Presentations/Clusters_5-6/34-FC-C6M11U1-Telehealth_and_Telemedicine.pdf
19. Videoconferencing system, Telemedicine. Available via https://www.who.int/medical_devices/innovation/videoconferencing_telemedicine.pdf (2011)
20. Triunfo, R., et al.: COTS technologies for telemedicine applications. Int. J. Comput. Assist. Radiol. Surg. **5**(1), 11–18 (2010)
21. Triunfo, R., et al.: Real-time telemedicine in pediatric cardiology. eTelemed2013 - The Fifth International Conference on eHealth, Telemedicine, and Social Medicine, pp. 320–326 (2013)
22. Zennaro, F., Neri, E., et al.: Real-time tele-mentored low cost "point-of-care US" in the hands of paediatricians in the emergency department: diagnostic accuracy compared to expert radiologists. PLoS One. **11**(10), e0164539 (2016)
23. Del Rio, M. et al.: Augmented reality for supporting real time telementoring: an exploratory study applied to ultrasonography. In: Proceeding of the 2nd International Conference on Medical and Health Informatics, pp. 218–222 (2018)
24. Rosenberg, J., et al.: SIP: Session Initiation Protocol. Internet Engineering Task Force (IETF). https://tools.ietf.org/html/rfc3261
25. Hardt, D.: The OAuth 2.0 Authorization Framework. Internet Engineering Task Force (IETF). Available via https://tools.ietf.org/html/rfc5849. 11(10)
26. Saxena, P.: Geo-location based augmented reality application. IJRET: Int J Res Eng Technol. **4**(7), 495–498 (2015)
27. Pagani, A., Henriques, J., Stricker, D.: Sensors for location-based augmented reality the example of galileo and egnos. ISPRS - International Archives of the Photogrammetry, Remote Sensing and Spatial Information Sciences. **418**(B1), 1173–1177 (2016)
28. Katiyar, A., Kalra, K., Garg, C.: Marker Based Augmented Reality. Advances in Computer Science and Information Technology (ACSIT). **2**(5), 441–445 (2015)
29. Reller, M.D., Strickland, M.J., Riehle-Colarusso, T., Mahle, W.T., Correa, A.: Prevalence of congenital heart defects in metropolitan Atlanta, 1998-2005. J. Pediatr. **153**(6), 807–813 (2008)
30. Wu, M.H., Chen, H.C., Lu, C.W., Wang, J.K., Huang, S.C., Huang, S.K.: Prevalence of congenital heart disease at live birth in Taiwan. J. Pediatr. **156**(5), 782–785 (2010)
31. Øyen, N., Poulsen, G., Boyd, H.A., Wohlfahrt, J., Jensen, P.K.A., Melbye, M.: Recurrence of congenital heart defects in families. Circulation. **120**(4), 295–301 (2009)
32. Wren, C., Reinhardt, Z., Khawaja, K.: Twenty-year trends in diagnosis of life-threatening neonatal cardiovascular malformations. Arch. Dis. Child. Fetal Neonatal Ed. **93**(1), F33–F35 (2008)
33. Fleiss, J.L.: Measuring nominal scale agreement among many raters. Psychol. Bull. **76**(5), 378–382 (1971)
34. Frexia, F., et al.: Preliminary cost-benefit analysis of a real-time telemedicine system. In: eTelemed2014 - The Sixth International Conference on eHealth, Telemedicine, and Social Medicine, pp. 257–262 (2014)

35. Sicotte, C., Lehoux, P., Van Doesburg, N., Cardinal, G., Leblanc, Y.: A cost-effectiveness analysis of interactive paediatric telecardiology. J. Telemed. Telecare. **10**(2), 78–83 (2004)
36. Philips Lumify: https://www.philips.it/healthcare/solutions/ultrasound
37. Adechotech: https://www.adechotech.com/products/
38. Lifesize: https://www.lifesize.com/en/video-conferencing-app/pricing
39. Caffery, L.J., Armfield, N.R., Smith, A.C.: Radiological interpretation of images displayed on tablet computers: a systematic review. Br. J. Radiol. **88**(1050), 20150191 (2015)
40. Boissin, C., Blom, L., Wallis, L., Laflamme, L.: Image-based teleconsultation using smartphones or tablets: qualitative assessment of medical experts. Emerg. Med. J. **34**(2), 95–99 (2017)
41. Raposo, V.L.: Telemedicine: The legal framework (or the lack of it) in Europe. GMS Health Technol Assess. **12**, Doc03 (2016)

Chapter 3
Innovative Technologies for Intrauterine Monitoring of Predictive Markers of Vascular and Neurological Well-Being

Silvia Visentin, Chiara Palermo, and Erich Cosmi

Contents

3.1 Developmental Origins Hypothesis

Intrauterine growth restriction (IUGR) is defined as an estimated fetal weight or an abdominal circumference below the 3th or 10th percentile for gestational age, in the base of a normal or abnormal fetal or maternal Doppler [1]. The term IUGR presents a prevalence of 1–3% and is correlated to higher mortality and morbidity than appropriate for gestational age fetuses (AGA) [2, 3].

Large epidemiologic studies have long suggested a strong correlation between IUGR and increased cardiovascular and metabolic events in adulthood. In particular, a new "developmental" model (Barker's theory) postulates that people, who

S. Visentin (✉) · E. Cosmi
Obstetric and Gynecological Unit, Department of Woman's and Child's Health,
University of Padua, Padua, Italy
e-mail: silvia.visentin.1@unipd.it; erich.cosmi@unipd.it

C. Palermo
Department of Cardiac, Thoracic and Vascular Sciences, University of Padua, Padua, Italy
e-mail: chiara.palermo@unipd.it

© Springer Nature Switzerland AG 2021
D. Pani et al. (eds.), *Innovative Technologies and Signal Processing in Perinatal Medicine*, https://doi.org/10.1007/978-3-030-54403-4_3

develop chronic disease, including coronary heart disease, stroke, high blood pressure, and type 2 diabetes, grow differently from other people during fetal life and childhood [4]. This could be connected to the concept of developmental plasticity, a critical window during development when a system is plastic and sensitive to nutritional, hormonal, and metabolic environment. For most organs and systems, the critical period occurs in utero and may give rise to a range of different physiological or morphological states in response to different conditions during development [5]. The ability of a human mother to nourish her fetus is partly determined by her own experience in utero, and her childhood growth, so the human fetus responds not only to conditions at the time of the pregnancy, but also occurring potentially several decades before [6]. In presence of a malnutrition, the fetus responds through some adaptations, increasing allocation of energy to the development of brain, heart, and adrenal glands, reducing blood flow in other organs, and producing lifelong changes in blood pressure and metabolism [9]. Well-established animal models have shown this mechanism by induced hypoxic-ischemia, maternal diabetes, and fetal exposure to glucocorticoids [7]. Furthermore, the "thrifty phenotype hypothesis", a constellation of metabolic and vascular fetus adaptations, could help to explain the link between IUGR and obesity, hypertension, osteopenia, diabetes, and cardiovascular disease (CVD). A central role seems to be played by insulin resistance.

Low birth weight (LBW) infants are known to have adipocytes that demonstrate increased numbers of insulin receptors, glucose uptake, and basal and insulin-stimulated insulin receptor substrate 1-associated phosphatidylinositol 3-kinase. Moreover, these adipocytes are resistant to insulin, resulting in a state of anti-lipolysis. An increased central adiposity is associated with increased free fatty acids, which stimulate the production of cholesterol and glucose, which in turn decrease insulin sensitivity. As fat deposition progresses in the liver, permanent structural hepatic changes occur. This molecular "switch" in insulin signaling "protects" the LBW fetus and neonate by promoting the storage of fat [8]. Finally, the "catch up growth theory", an undernutrition in utero followed by rapid childhood growth, seems to affect the onset of later diseases. Developing a high body mass during childhood, these children may have a disproportionately high-fat mass in relation to lean body mass, which will lead to insulin resistance, establishing a back and forward circle [9].

3.2   Impact of CVD

CVD is the leading cause of mortality, morbidity, and hospitalization, in both genders, and accounted for over 17,9 millions of deaths worldwide [WHO 2017]. CVD evolves gradually and may interfere with quality of life, physical disability, and lifelong dependence on health services and medications. Diabetes type II is recognized as an independent risk factor for CVD even when under glycemic control, and endothelial cell (EC) dysfunction is associated with both diabetes and the pathogenesis of CVD [10]. Moreover, in the USA, obesity has more than doubled among children and adolescents increasing actually from 5% to 17.6% [*Atlanta GA, Centers for Disease Control and Prevention. Prevalence of Obesity Among U.S. Children and Adolescents (Aged 2–19 Years) National Health and Nutrition Examination Surveys, NHANES (1976–1980 and 2003–2006). 2009*] and hypertension in childhood is a strong risk factor for later cardiovascular disease and is considered an indication for lifestyle modifications. Establishing the mechanisms linking these factors, already in the prenatal period, it could provide essential insights and inform novel therapeutics. An appropriate perinatal selection of IUGR fetuses at risk of future CVD (gestational age at onset or fetoplacental Doppler changes) would allow an efficient approach to detect those cases who may later benefit from early screening and intervention in infancy [11].

The most important cardiovascular markers in the neonatal or pediatric age are some fetal echocardiographic measures (diastolic function, strain parameter), mean blood pressure, endothelial function, urinary proteinuria, weight gain, and the proportion of body fat mass.

3.3   Endotelial Dysfunction

Vascular endothelial cells line the inner surface of blood vessels and act as a metabolically active monolayer, which is constantly exposed to both biochemical and biomechanical stimuli. It is well established that the transduction of these stimuli, alone or in combination, by the endothelium determines the physiology or pathology of the cardiovascular system. As a major regulator of local vascular homeostasis, the endothelium maintains the balance between vasodilatation and vasoconstriction, inhibition and promotion of the proliferation and migration of smooth muscle cells, prevention and stimulation of the adhesion and aggregation of platelets, as well as thrombogenesis and fibrinolysis [12]. The term "endothelial dysfunction" was coined by Furchgott and Zawadzki who discovered that acetylcholine requires the presence of the endothelial cells to relax the underlying vascular smooth muscle. Mediators responsible for dilatator mechanisms include agents such as nitric oxide (NO), prostacyclin, and endothelium-derived hyperpolarizing factor substances [13].

Given the vast range of vasoprotective effects of NO, the term endothelial dysfunction generally refers to reduce NO bioavailability, through decreased eNOS expression, cause of an enhanced vasoconstrictor responses, and an impaired endothelium-dependent vasodilation, respectively.

Although endothelial dysfunction occurs in many different disease processes, oxidative stress can be identified as a common denominator. ROS play a central role in vascular physiology and their overproduction is of particular relevance to vascular pathologies [14]. Endothelial cells in areas of high shear stress have shown an increase in lipid uptake, monocyte adhesion, apoptotic rates, and revealed distinctive patterns of upregulated gene expression not previously observed in EC, such as the endothelial isoform of nitric oxide synthase (ecNOS), the inducible isoform of cyclooxygenase (COX-2), and manganese-dependent superoxide dismutase. These enzymes can exert potent antithrombotic, anti-adhesive, anti-proliferative, anti-inflammatory, and antioxidant effects both within the endothelial lining and in interacting cells, such as platelets, leukocytes, and smooth muscle ("athero-protective gene hypothesis") [15].

The mechanisms that relate fetal programming to LBW and endothelial dysfunction in utero are still largely unknown and different explanations have been proposed [9]. Many animal studies have described the presence of vascular structural alterations in the offspring of protein and caloric restricted dams, such as remodeling of aorta and mesenteric arteries, capillary rarefaction, increased arterial stiffness, alterations in the composition and structure of the extracellular matrix of the vessels, decreased angiogenesis and an increase of blood pressure [16]. Moreover, the role of glucocorticoids in the fetal programming seems to be important; in the condition of placenta insufficiency there is a decreased activity of 11b-hydroxysteroid dehydrogenase-2, which metabolizes in the placenta corticosterone to the inert 11-dehydrocorticosterone, which would in turn increases access of endogenous maternal cortisone to the fetus and leads to the LBW and elevate blood pressure into adult life [17].

Alterations in maternal lipid profile may also be involved in the process. Aortas from spontaneously aborted fetuses and from died children aged 1–13 years of hypercholesterolemic mothers contained significantly more and larger fatty streaks than those of normocholesterolemic mothers, demonstrating an involvement of lipid peroxidation and the influence of the maternal environment over the onset of atherosclerosis. Oxidized low-density lipoproteins (OxLDL) interference with intracellular signaling pathways regulating the expression of many genes that determine the recruitment of cells, their proliferation and differentiation, metabolic and secretory activity, and death. OxLDL is rapidly internalized by macrophages within atherosclerotic lesions, which leads to foam cell formation, a key event in fatty streak formation, and also promotes further LDL oxidation. An imbalance between oxidants/antioxidants resulting in higher oxidative stress has been suggested as one of the mechanisms responsible for the detrimental effect of high-fat diet [18]. Figure 3.1 summarizes it.

The fetal insulin hypothesis offers an alternative explanation for the consistent association between impaired fetal growth, insulin resistance, and the link with hypertension and vascular disease. Monogenic diseases determined insulin resistance could result in low insulin-mediated fetal growth in utero as well as insulin resistance in childhood and adulthood. Angiogenesis could be impaired in insulin-resistant fetal tissues where the generation of nitric oxide might be deficient, resulting in impaired vasodilation, hence decreased blood flow, a poorly developed capillary circulation in vulnerable organs and a deficient endothelium-dependent vasodilation [19].

Fig. 3.1 Endothelial pathways in the balance between vasodilatation and vasoconstriction, inhibition and promotion of the proliferation and migration of smooth muscle cells, prevention and stimulation of the adhesion and aggregation of platelets, as well as thrombogenesis and fibrinolysis

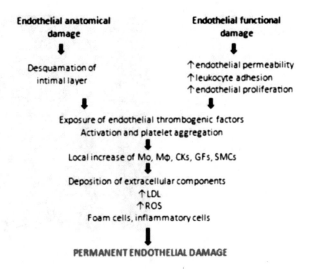

3.4   Vascular Function Evaluation

It is well known that atherogenesis begins in early life. Autopsies on children (2–15 years of age) who died from circumstances unrelated to CVD report fatty streaks and fibrous plaque lesions in the aorta, suggesting arterial wall damage may begin during childhood [20]. Recently, the improvement of imaging in health care allowed studying and analyzing the structure, thickness, and functionality of several vessels. It is well known that carotid intima-media thickness (cIMT) and arterial stiffness are clinical indicators of atherosclerosis and increased CVD in the general adult population and in at-risk individuals [21]. More recently, ultrasound-based measurement of aorta intima-media thickness (aIMT) in children has become a feasible, accurate, and sensitive marker of atherosclerosis risk [22].

The measure of aIMT was from 2009 possible also in utero. Cosmi et al. published an article, in which the higher ultrasonographic resolution was able to assess early vascular changes that may be linked to atherosclerosis. Early endothelial dysfunction, as an impairment of arterial vasodilatory function, may play an important role in premature stiffening of the aortic vessels, which predisposes these individuals to hypertension, stroke, nephropathies, and metabolic syndrome. Infants who had IUGR, as Skilton demonstrated, have at birth a thicker aorta, suggesting that prenatal events (e.g., impaired fetal growth) might be associated with structural changes, in the main vessels, which probably highlighted in utero.

IMT and diameter were measured in a coronal or sagittal view of the fetus at the dorsal arterial wall of the most distal 15 mm of the abdominal aorta, sampled below the renal arteries and above the iliac arteries, as previously described [23]; gain settings were used to optimize image quality. aIMT was defined as the distance between the leading edge of the blood–intima interface and the leading edge of the media-adventitia interface on the far wall of the vessel [24]. This vessel was selected

Fig. 3.2 Fetal aIMT
measurement

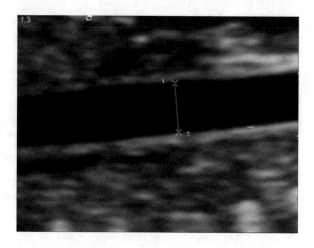

because it is reported to be the first involved in the atherosclerotic process, in particular the dorsal arterial wall, which is the most lesion-prone site seen in autopsy samples. Three measurements were taken, and the arithmetic mean aortic intima-media thickness was considered for the study (Fig. 3.2). Aortic diameter was measured at the same level of aortic intima-media thickness, from the inner wall to the wall edges.

Furthermore, Doppler studies of the fetal circulation in IUGR and hypoxia have demonstrated increased resistance to flow in the umbilical arteries and redistribution in the fetal circulation, with a decreased afterload at the level of the left ventricle due to the brain-sparing effect, resulting in reduced cerebral impedance and increased resistance in the descending aorta [25].

In mammals, the large arteries, such as the human aorta, provide an important energy-storage and pulse-dampening function. The energy from the pressure of ejecting blood during systole is stored as strain energy in the distended artery wall. This strain is returned during diastole when the arteries return to their original dimensions. During diastole the aortic valve is closed, preventing blood from returning to the left ventricle and instead allowing the blood flow to move further down the arterial tree with a dampened pulse wave. The major component is the elastin protein that provides the elasticity necessary for cyclic deformation of the arterial wall and is mainly located in the tunica media of the arterial wall [26].

In the smooth muscle cells of the arterial wall that are responsible for the extracellular matrix composition, elastin expression occurs over a short period during development from mid-gestation and continuing through the postnatal period at high levels [27, 28]. However, the expression of elastin in the aorta decreases rapidly when the physiological rise in blood pressure stabilizes postnatally, and there is minimal elastin synthesis in the adults. This explains why the repair of elastic fibers is incomplete and elastin protein has a long half-life in adults [27]. The extracellular matrix in great vessel walls in mammals has multiple functions, i.e., providing mechanical and structural properties and instructional signals, that control vascular cell phenotypes and function (e.g., influencing smooth muscle cell gene expression). This reciprocal interaction between the extracellular matrix and smooth muscle cells of the arterial wall is of paramount importance for directing the developmental transitions that occur in embryogenesis, in postnatal development, and in response to injury [27].

In absence of the possibility to know the intravascular fetal pressure, the study of the aortic fetal diameter permits to investigate the vessel's stiffness, a possible marker of aortic atherosclerosis. The diameter is measured at the same level as the aIMT, from the inner wall to the wall edges, taking at a maximal systolic and a minimal diastolic diameter. It is measured using the cine-loop capability of the ultrasound machine, once the images of the entire cardiac cycle were frozen. The vessel was visualized in a longitudinal view of the fetus. The transducer was tilted to obtain an angle of insonation as close to 0° as possible and always less than 30°; the high-pass filter was reduced to the minimum. Each measurement was taken during fetal apnea after three consecutive, similar waveforms were obtained (Fig. 3.3).

The observations [29] show that the change between the systolic and diastolic fetal abdominal aorta diameters was significantly greater in IUGR fetuses than in controls, as the change between the systolic and diastolic velocity. An important function of the human aorta is to store strain energy in the distended aortic wall during systole. The main component of strain energy storage is the extracellular matrix of the tunica media. The composition of the extracellular matrix depends on different stimuli such as blood pressure and flow that could both be modified by increased peripheral resistance to flow. In the IUGR fetuses, an increased aIMT could reflect a different extracellular matrix composition, as a different aorta compliance and distensibility. A significant increase in systolic and diastolic abdominal aorta diameters in IUGR fetuses that could reflect a compensatory mechanism secondary to several changes at the level of fetal aorta involving, among other things, increased peripheral resistances associated with brain-sparing mechanisms. Generally, vessels with distensible elasticity present a universal vessel wall elastic modulus that applies across species and in vessels with a different extracellular matrix composition. This

Fig. 3.3 Fetal aortic diameter measurement

suggests strong evolutionary pressure to ensure that all elastic arteries have similar mechanical properties at each organism's mean physiological blood pressure. This universal elastic modulus probably reflects a target mechanical property that is best able to operate in a pulsatile circulatory system [27]. There is evidence in the literature that organisms are able, especially during fetal life, to adjust the mix of extracellular matrix components in the vessel wall to produce the mechanical properties appropriate for different hemodynamic variables to achieve the universal elastic modulus and IUGR fetuses could be one of these examples.

3.5   Cardiac Function

During pregnancy, there are many physiological changes due to the increased metabolic demand of the mother-fetus couple, which requires an adequate utero-placental circulation. Impairment of these mechanisms of adaptation can cause a fetal or maternal disease, such as growth retardation and preeclampsia, or unmask an underlying cardiac disease. Pregnancy is associated with an increase in heart rate, which starts in the first trimester, peaks in the third trimester (15–25% increase over the baseline heart rate), and returns to pre-conceptional values by 10 days postpartum [30].

Echocardiography is the most frequently used imaging technique to assess cardiac function and hemodynamic. It allows a rapid assessment of systolic and diastolic function of cardiac chambers, regional wall motion and valve anatomy and function. Due to the safety of ultrasounds, the wide availability of the technique, and its portability and repeatability, echocardiography is very useful to assess the cardiovascular system of pregnant women with suspected or confirmed heart disease [31]. This technique is actually used also in prenatal diagnosis to investigate the myocardial function, especially in the presence of intrauterine growth restriction. Several are the methods used.

Deformation imaging is an echocardiographic technique used to assess myocardial function by measuring the actual change in length of the myocardium through the cardiac cycle. Myocardial deformation may be evaluated through two methods: the first, tissue velocity imaging (TDI), is a Doppler-based method, whereas the second, speckle-tracking echocardiography (STE), is based on the analysis of conventional two-dimensional grayscale images. Strain and strain rate, indices of myocardial deformation, can be obtained with both TDI and STE. Since TDI is a Doppler-based method and velocity can only be measured along the direction of the ultrasound beam, only a limited number of strain components can be measured by TDI [32]. Conversely, STE is based on the detection of 2D images of the motion of acoustic markers (called "speckles") generated by the interaction of ultrasounds with the myocardium. The position of the speckles can be tracked during the cardiac cycle by using specific software packages. The movement of speckles can be used to measure strain and calculate strain rate. To analyze the different components of myocardial deformation (strain) with STE it is necessary to acquire several views: 4-chamber, 2-chamber, and apical long-axis views to compute global longitudinal strain (GLS) and short-axis views for circumferential and radial strains [33].

Moreover, since it measures directly myocardial function, deformation imaging can detect subclinical myocardial dysfunction, when ejection fraction or other chamber function parameters are still in the normal range because the heart has activated its compensatory mechanisms. Since, during pregnancy, there is a continuous variation of the loading conditions of the heart, the use of STE can be particularly useful to study the changes occurring in the myocardial function during either normal or pathological pregnancy (Figs. 3.4 and 3.5).

To obtain images to be analyzed with the STE software package, it is recommended to optimize image quality by using a grayscale second-harmonic 2D imaging technique with careful adjustment of image contrast. The gain settings should be optimized, the depth should be reduced, and the focus should be in the middle of the left ventricle. Finally, images should have an adequate temporal resolution (50–90 frames per second). Lower temporal resolutions will not allow a sufficient number of systolic frames to track the motion of the kernels. Higher temporal resolution will impact the spatial resolution of the images by reducing the number of scan lines [34]. Moreover, it is essential to optimize the left ventricle border visualization. Care must be taken to avoid left ventricle foreshortening and image acquisition should be performed during breath-hold to minimize respiratory interference. It is essential that the electrocardiographic trace is stable to avoid artifacts during the evaluation and at least three cardiac cycles should be acquired for each loop. Artifacts, such as reverberation or shadowing, could affect strain computation and provide wrong strain values, which might erroneously suggest cardiac dysfunction [35] (Fig. 3.6).

Fig. 3.4 The image shows the three main components of myocardial deformation: longitudinal, radial, and circumferential

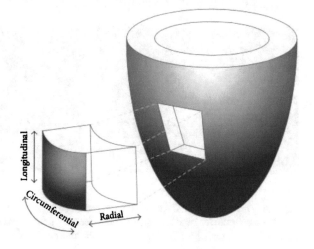

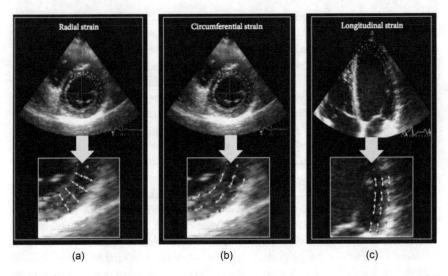

Fig. 3.5 The figure shows speckle-tracking analysis and the main components of myocardial deformation: radial (**a**), circumferential (**b**), and longitudinal (**c**)

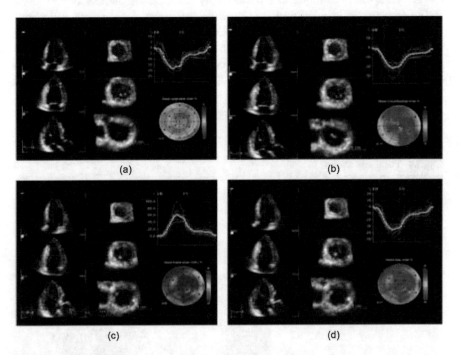

Fig. 3.6 Three-dimensional speckle-tracking echocardiography. The software package calculates simultaneously the longitudinal (**a**), the circumferential (**b**), and the radial (**c**) components of myocardial deformation, plus a composite parameter (area strain) (**d**)

3.6   Clinical Applications of Deformation Imaging in Pathological Pregnancies

3.6.1   Maternal Cardiac Study in Preeclampsia

Preeclampsia is a hypertensive complication that affects 5–7% of pregnancies and is one of the most common causes of maternal morbidity and mortality [36]. In fact, it is considered a complex multiorgan disease potentially involving the kidney, liver, and cardiovascular and hematologic systems, as well as the brain. Autopsy data have shown a tenfold prevalence of myocardial contraction band necrosis in pre-eclamptic patients if compared with pregnant women that died from other causes [37]. Some studies have demonstrated persistent maternal cardiac impairment and hemodynamic changes years after delivery. Valensise et al. demonstrated that signs of left ventricular diastolic dysfunction and persistent heart remodeling persist in non-pregnant women before a second pregnancy with recurrent preeclampsia. These findings could raise an issue concerning preeclampsia as a cause or effect of heart remodeling [38, 39].

The majority of published papers about echocardiographic assessment of maternal heart in preeclamptic women have used conventional parameters of cardiac function and remodeling [40, 41]. In preeclamptic patients, the use of TDI and 2D STE demonstrated a reduction of both left and right ventricular diastolic and systolic function, also in preclinical stages of the disease, when cardiac output and EF are still preserved. LV-mass, LV-mass index, and LV wall thickness in preeclamptic women are higher than in healthy controls, reflecting the increase in LV afterload. Myocardial performance index (MPI), an index of reduced cardiac systolic and diastolic function, also increased. An impairment of right ventricular systolic function has been also described, reflecting the increase of pulmonary resistance secondary to LV diastolic dysfunction [42].

Moreover, several authors agree that longitudinal, radial, and circumferential strains are impaired in preeclampsia and may remain impaired also for months after the delivery, even in patients with preserved cardiac output and ejection fraction. An interesting consideration is that coexisting left ventricle hypertrophy and regional longitudinal systolic dysfunction could reflect a regional subendocardial impairment, probably due to subendocardial ischemia and/or fibrosis [43]. Currently, new parameters based on 3D STE are emerging, showing an increased ability to detect subclinical myocardial impairment and early systolic and diastolic cardiac dysfunction. Myocardial dysfunction precedes chamber impairment and 3D STE can provide an assessment of global and regional LV function. Furthermore, some authors, using 3D STE, demonstrated that early-onset preeclamptic patients presented worse cardiac remodeling than late-onset preeclamptic patients, underlining the clinical relevance of detecting earlier and subtler cardiac dysfunction signs. Other authors distinguish between preeclamptic and non-proteinuric hypertensive women, showing less impairment of longitudinal, circumferential, and radial strain in the latter. These findings could mean that hypertension may not be the only cause of preeclamptic heart impairment, although women with no proteinuric hypertension presented a worse cardiac function than healthy patients [44].

3.6.2 Maternal Cardiac Study in Intrauterine Growth Restriction Disease

Knowledge of heart disease in women is constantly evolving and emerging data show that complications of pregnancy such as preeclampsia and intrauterine growth restriction (IUGR) are predictors for the development of heart disease later in life. Up to 10% of all pregnancies are affected by IUGR and its definition is controversial. The main reasons of IUGR are placental insufficiency and defective trophoblastic invasion, currently evaluated by the estimated fetal weight and umbilical artery Doppler flow velocity. Fetal Doppler evaluation is a useful method to predict fetal compromise and permits distinguishing between severe IUGR and small for gestational age fetuses. However, different classifications are also reported in the literature and they could generate confusion in the definition of both maternal and fetal risk [45].

While maternal cardiac modifications occurring during normal pregnancy are well known, in normotensive IUGR pregnancies, there are contradictory lines of evidence about maternal hemodynamic. Some authors reported reduced cardiac output and left ventricular compliance, whereas others reported reduced maternal systolic cardiac function and increased total vascular resistance, without alterations of left diastolic function compared to physiological pregnancies [46, 47].

Moreover, IUGR patients, compared with preeclamptic pregnancies, seem to present lower cardiac index, left ventricular diastolic dysfunction, and higher total vascular resistance index. Unlike preeclampsia, cardiac geometry and intrinsic myocardial contractility were reported to be preserved, but a third of IUGR patients present reduced diastolic reserve and an overt diastolic chamber dysfunction, despite a normal ejection fraction [48]. This suggests that the cardiovascular response is similar to that seen in preeclamptic patients, though less severe. Lack of physiological adaptation to the pregnancy, assimilating IUGR patients to a nonpregnant hemodynamic condition, could explain the reason of high resistance, low blood volume, and hypotensive condition, which characterized IUGR patient's condition The introduction of TDI and 2D STE techniques for analysis of myocardial deformation might allow an earlier diagnosis and better grading of cardiac dysfunction [49]. While several authors described the feasibility of STE in studying fetal heart function and morphology, in particular, the segmental and global systolic and diastolic velocities, strain, and strain rate values, few studies described its application for the evaluation of IUGR patients. Krause et al. investigated maternal longitudinal mechanical dyssynchrony, a useful tool used for the evaluation of LV function, finding that pregnancies complicated by IUGR recorded significantly higher degrees of inter- and intraventricular dyssynchrony than those of normal controls [50].

Reduced maternal cardiac function in pregnancies that are complicated by preeclamptic and intrauterine growth restriction is the result of both reduced intrinsic myocardial contractility and reduced diastolic filling. Myocardial dysfunction can be present even in the presence of a normal ejection fraction, with significant decreases in radial, circumferential, and longitudinal strain values.

The use of 2D and 3D STE techniques to evaluate ventricular mechanics may help detect subclinical left ventricular dysfunction in women affected by obstetrical pathologies as preeclampsia and intrauterine growth restriction. Early detection of left ventricular dysfunction with the institution of appropriate treatment may reduce the risk of future CVD.

3.6.3 Fetal Cardiac Study in Intrauterine Growth Restriction Disease

The heart is a central organ in the fetal adaptive mechanisms to placental insufficiency and cardiac dysfunction in utero is recognized among the essential pathophysiologic features in the presence of intrauterine growth restriction. Usually, the cardiac fetal study considers the cardiac morphometry, the diastolic, and the systolic function. The cardiac geometry includes left and right sphericity index (defined as base-to-apex length divided by basal diameter) measured on an end-diastolic 2D apical 4-chamber view. Left ventricular end-diastolic diameter and septal and posterior wall thickness were measured by M-mode on a parasternal long-axis parasternal view and relative wall thickness, calculated as posterior wall plus interventricular septum thickness divided by left ventricular diameter.

Systolic function evaluation includes cardiac output, left ejection fraction, left ventricular thickening, mitral/tricuspid annular displacement (MAPSE/TAPSE), and systolic annular peak velocity. Diastolic function was evaluated by peak early and late transvalvular filling (E/A) ratio, deceleration time of E velocity, isovolumetric relaxation time (IRT), early (E=) and late (A=) diastolic annular peak velocities, and E/E = ratio.

The more recent literature shows that fetuses and children with IUGR present more globular hearts, decreased stroke volume compensated by an increased heart rate, subclinical systolic dysfunction, diastolic changes, and increased blood pressure and vascular wall thickness [51]. The postnatal globular cardiac shape is most likely the result of changes in cardiac development, induced by sustained pressure increase. Although in normal conditions the resulting increased wall stress on the developing myocardial fibers should trigger a compensating hypertrophic response, sustained hypoxia and undernutrition might render the myocardium unable to develop hypertrophic changes. In these circumstances, increased wall stress can be compensated for only by increasing the local radius of curvature, resulting in dilation and a more spherical cavity. The different types of intrauterine growth restriction (early and late, if below or not 32 weeks of gestation) could influence the cardiac remodeling response to intrauterine under-nutrition and hypoxia. In particular, late-onset IUGR, being exposed to a milder prenatal insult, can compensate compromised longitudinal fibers by increasing radial contractility. Longitudinal dysfunction compensated by increased radial function without evident myocardial hypertrophy has been described as a common feature of early-stage cardiac compromise in ischemic myocardial disease and hypertensive and athletes' hypertrophic cardiomyopathy [52]. In contrast, early IUGR cannot compensate similarly because the insult is more severe and needs to increase heart rate to maintain cardiac output. Indeed, endocardial longitudinal fibers are the farthest ones from the epicardial blood supply and consequently the most sensitive under milder degrees of hypoxia. Conversely, radial fibers constitute the middle layer of the myocardial wall and are affected in advanced stages of hypoxia. These findings add to the body of evidence indicating important pathophysiological differences in early- and late-onset IUGR, which may help to understand the differences in natural history and outcome.

3.7 Conclusions

After Barker's hypothesis, an increasing number of subsequent epidemiological studies confirmed the link among low birth weight, rapid weight gain in the first years of life, obesity in adolescent period, and increased risk of CVD, stroke, glucose intolerance, and type II diabetes in adult life [53]. Studies conducted in fetuses, neonates, children, and adolescents born IUGR point to the possibility that endothelial dysfunction, evaluated by aIMT, cIMT, carotid stiffness, central pulse wave velocity, brachial artery flow-mediated dilation, endothelium-dependent microvascular vasodilatation, echocardiographic evaluation, and arterial blood pressure, may be an inborn characteristic of subjects with LBW that persists in childhood into adult life [22, 54–56]. Already in 1997, one of the largest cohort studies of almost 150.000 adolescents in Sweden showed that systolic blood pressure was significantly higher in young men who had the lowest birth weight, thus supporting the notion of a programming effect of IUGR in utero on hemodynamic regulation in early adult life [57].

IUGR is a very complex and multifactorial disorder with long-term persistence of CVD older in patients who suffered IUGR early in life. Whereas the dominant focus of experimental studies to date has been on defining the phenotypic consequences of perturbations of maternal nutrition, the emphasis has now shifted to determining those initiating mechanisms through which early nutrition and associated growth patterns result in cardiovascular and metabolic dysfunction. There is a clear requirement to reconcile the balance of contribution of the "thrifty phenotypes" and "thrifty genotypes", investigating by experimental and epidemiological studies the impact of relative under and over-nutrition in prenatal life. Molecular and epigenetic mechanisms of programming, after different initial insults through animal models and in vitro techniques, might consent to understand common mechanisms leading to CVD, preventing or alleviating CVD and renal effects of programming in adulthood. Innovative interdisciplinary research in the areas of nutrition, reproductive physiology, and vascular biology will play an important role in developing new strategies of intervention. Future pharmaceutical therapies and innovative alternative nutritional/environmental strategies even could maximize population-based well-being.

The prenatal and postnatal vascular modifications showed in several human and animal studies by blood pressure and cardiac modifications in IUGR, which appear small, seem to be clinically relevant and/or a major contributor to Barker's hypothesis association of birth weight with ischemic heart disease. Whether the American Heart Association and the European guidelines on CVD prevention in clinical practice encourage the clinical follow-up of patients with a high risk of CVD, current clinical pediatric guidelines do not include IUGR as a risk factor and several studies are still needed to understand its role, the importance of long-term cardiovascular follow-up and the correct lifestyle of patients who suffered early or late IUGR in utero [11].

References

1. Gordijn, S.J., et al.: Consensus definition of fetal growth restriction: a Delphi procedure. Ultrasound Obstet. Gynecol. **48**, 333–339 (2016)
2. Lausman, A., Kingdom, J., Gagnon, R., Basso, M., Bos, H., Crane, J., Davies, G., Delisle, M.F., Hudon, L., Menticoglou, S., Mundle, W., Ouellet, A., Pressey, T., Pylypjuk, C., Roggensack, A., Sanderson, F., Maternal Fetal Medicine Committee: Intrauterine growth restriction: screening, diagnosis, and management. J. Obstet. Gynaecol. Can. **35**, 741–757 (2013)
3. American College of Obstetricians and Gynecologists: ACOG Practice bulletin no. 134: fetal growth restriction. Obstet. Gynecol. **121**, 1122–1133 (2013)
4. Barker, D.J., Osmond, C., Golding, J., Kuh, D., Wadsworth, M.E.: Growth in utero, blood pressure in childhood and adult life, and mortality from cardiovascular disease. BMJ. **298**, 564–567 (1989)
5. Barker, D.J., Winter, P.D., Osmond, C., Margetts, B., Simmonds, S.J.: Weight in infancy and death from ischaemic heart disease. Lancet. **2**, 577–580 (1989)
6. Mellanby, E.: Nutrition and child-bearing. Lancet. **222**(5751), 1131–1137 (1933)
7. Harding, J.E.: The nutritional basis of the fetal origins of adult disease. Int. J. Epidemiol. **30**, 15–23 (2001)
8. Nobili, V., Marcellini, M., Marchesini, G., Vanni, E., Manco, M., Villani, A., et al.: Intrauterine growth retardation, insulin resistance, and non-alcoholic fatty liver disease in children. Diabetes Care. **30**, 2638–2640 (2007)
9. Barker, D.J., Eriksson, J.G., Forsen, T., Osmond, C.: Fetal origins of adult disease: strength of effects and biological basis. Int. J. Epidemiol. **31**, 1235e9 (2002)
10. Wilmot, E., Edwardson, C., Achana, F., et al.: Sedentary time in adults and the association with diabetes, cardiovascular disease and death: systematic review and meta-analysis. Diabetologia. **55**, 2895–2905 (2012)
11. Kavey, R.E., Allada, V., Daniels, S.R., et al.: Cardiovascular risk reduction in high-risk pediatric patients: a scientific statement from the American Heart Association Expert Panel on Population and Prevention Science; the Councils on Cardiovascular Disease in the Young, Epidemiology and Prevention, Nutrition, Physical Activity and Metabolism, High Blood Pressure Research, Cardiovascular Nursing, and the Kidney in Heart Disease; and the Interdisciplinary Working Group on Quality of Care and Outcomes Research: endorsed by the American Academy of Pediatrics. Circulation. **114**, 2710–2738 (2006)

12. Martin, F.A., Murphy, R.P., Cummins, P.M.: Thrombomodulin and the vascular endothelium: insights into functional, regulatory, and therapeutic aspects. Am. J. Physiol. Heart Circ. Physiol. **304**, H1585–H1597 (2013)

13. Flavahan, N.A., Vanhoutte, P.M.: Endothelial cell signaling and endothelial dysfunction. Am. J. Hypertens. **8**, 28S–41S (1995)

14. Griendling, K.K., Fitz Gerald, G.A.: Oxidative stress and cardiovascular injury. Part I. Basic mechanisms and in vivo monitoring of ROS. Circulation. **108**, 1912–1916 (2003)

15. Mover and Shakers in the Vascular Treed: Hemodynamic and Biomechanical Factors in Blood Vessel Pathology, Mar 11–12, 1999, in Bethesda

16. Nuyt, A.M.: Mechanisms underlying developmental programming of elevated blood pressure and vascular dysfunction: evidence from human studies and experimental animal models. Clin. Sci. Lond. **114**, 1–17 (2008)

17. Langley-Evans, S.C., Phillips, G.J., Benediktsson, R., Gardner, D.S., Edwards, C.R., Jackson, A.A., Seckl, J.R.: Protein intake in pregnancy, placental glucocorticoid metabolism and the programming of hypertension in the rat. Placenta. **173**, 169–172 (1996)

18. Berliner, J.A., Navab, M., Fogelman, A.M., et al.: Atherosclerosis: basic mechanisms. Oxidation, inflammation, and genetics. Circulation. **91**, 2488–2496 (1995)

19. Hattersley Andrew, T., Tooke, J.E.: The fetal insulin hypothesis: an alternative explanation of the association of low birth weight with diabetes and vascular disease. Lancet. **353**, 1789–1792 (1999)

20. McGill Jr., H.C., McMahan, C.A., Herderick, E.E., Malcom, G.T., Tracy, R.E., Strong, J.P.: Origin of atherosclerosis in childhood and adolescence. Am. J. Clin. Nutr. **72**, 1307S–1315S (2000)

21. Lorenz, M.W., Markus, H.S., Bots, M.L., Rosvall, M., Sitzer, M.: Prediction of clinical cardiovascular events with carotid intima-media thickness: a systematic review and meta-analysis. Circulation. **115**, 459–467 (2007)

22. Skilton, M.R., Evans, N., Griffiths, K.A., Harmer, J.A., Celermajer, D.S.: Aortic wall thickness in newborns with intrauterine growth restriction. Lancet. **365**, 1484–1486 (2005)

23. Koklu, E., Ozturk, M.A., Gunes, T., Akcakus, M., Kurtoglu, S.: Is increased intima-media thickness associated with preatherosclerotic changes in intrauterine growth restricted newborns? Acta. Paediatr. **96**, 1858 (2007)

24. Koklu, E., Kurtoglu, S., Akcakus, M., Yikilmaz, A., Coskun, A., Gunes, T.: Intima-media thickness of the abdominal aorta of neonates with different gestational ages. J. Clin. Ultrasound. **35**, 491–497 (2007)

25. Hecher, K., Campbell, S., Doyle, P., et al.: Assessment of fetal compromise by Doppler ultrasound investigation of the fetal circulation. Arterial, intracardiac, and venous blood flow velocity studies. Circulation. **91**, 129–138 (1995)

26. Cheng, J.K., et al.: Extracellular matrix and the mechanics of the large artery development. Biomech. Model. Mechanobiol. **11**, 1169–1186 (2012)

27. Wagenseil, J.E., Mecham, R.P.: Vascular extracellular matrix and arterial mechanics. Physiol. Rev. **89**, 957–989 (2009)

28. Bendeck, M.P., Keeley, F.W., Langille, B.L.: Perinatal accumulation of arterial wall constituents: relation to hemodynamic changes at birth. Am. J. Phys. **267**, H2268–H2279 (1994)

29. Visentin, S., Londero, A.P., et al.: Fetal abdominal aorta: doppler and structural evaluation of endothelial function in intrauterine growth restriction and controls. Ultraschall Med. **40**(1), 55–63 (2019)

30. Sanghavi, M., Rutherford, J.D.: Cardiovascular physiology of pregnancy. Circulation. **130**(12), 1003–1008 (2014)

31. Liu, S., Elkayam, U., Naqvi, T.Z.: Echocardiography in pregnancy: part 1. Curr. Cardiol. Rep. **18**(9), 92 (2016)

32. Hoit, B.D.: Strain and strain rate echocardiography and coronary artery disease. Circ. Cardiovasc. Imaging. **4**(2), 179–190 (2011)

33. Blessberger, H., Binder, T.: Two dimensional speckle tracking echocardiography: clinical applications. Heart. **96**(24), 2032–2040 (2010)
34. Teske, A.J., De Boeck, B.W.L., Melman, P.G., Sieswerda, G.T., Doevendans, P.A., Cramer, M.J.M.: Echocardiographic quantification of myocardial function using tissue deformation imaging, a guide to image acquisition and analysis using tissue Doppler and speckle tracking. Cardiovasc. Ultras. **5**, 27 (2007)
35. Voigt, J.U., Pedrizzetti, G., Lysyansky, P., et al.: Definitions for a common standard for 2D speckle tracking echocardiography: consensus document of the EACVI/ASE/Industry Task Force to standardize deformation imaging. Eur. Heart J. Cardiovasc. Imaging. **16**(1), 1–11 (2015)
36. Walker, J.J.: Pre-eclampsia. Lancet. **356**, 1260–1265 (2000)
37. Bauer, T.W., Moore, G.W., Hutchins, G.M.: Morphologic evidence for coronary artery spasm in eclampsia. Circulation. **65**(2), 255–259 (1982)
38. Valensise, H., Lo Presti, D., Gagliardi, G., et al.: Persistent maternal cardiac dysfunction after preeclampsia identifies patients at risk for recurrent preeclampsia. Hypertension. **67**(4), 748–753 (2016)
39. Orabona, R., Vizzardi, E., Sciatti, E., et al.: Insights into cardio-vascular alterations after pre-eclampsia: a 2D strain echocardiographic study. Eur. J. Heart Fail. **18**, 391 (2016)
40. Hamad, R.R., et al.: Assessment of left ventricular structure and function in preeclampsia by echocardiography and cardiovascular biomarkers. J. Hypertens. **27**(11), 2257–2264 (2009)
41. Bamfo, J.E.A.K., et al.: Maternal cardiac function in normotensive and pre-eclamptic intra-uterine growth restriction. Ultrasound Obstet. Gynecol. **32**(5), 682–686 (2008)
42. Fayers, S., Moodley, J., Naidoo, D.P.: Cardiovascular haemodynamics in pre-eclampsia using brain natriuretic peptide and tissue Doppler studies. Cardiovasc. J. Afr. **24**(4), 130–136 (2013)
43. Melchiorre, K., Sutherland, G.R., Baltabaeva, A., Liberati, M., Thilaganathan, B.: Maternal cardiac dysfunction and remodeling in women with preeclampsia at term. Hypertension. **57**(1), 85–93 (2011)
44. Cong, J., Fan, T., et al.: Maternal cardiac remodeling and dysfunction in preeclampsia: a three-dimensional speckle-tracking echocardiography study. Int. J. Cardiovasc. Imaging. **31**(7), 1361–1368 (2015)
45. Unterscheider, J., Daly, S., Geary, M.P., et al.: Optimizing the definition of intrauterine growth restriction: the multicenter prospective PORTO Study. Am. J. Obstet. Gynecol. **208**(4), 290.e1–290.e6 (2013)
46. Vasapollo, B., Valensise, H., Novelli, G.P., et al.: Abnormal maternal cardiac function and morphology in pregnancies complicated by intrauterine fetal growth restriction. Ultrasound Obstet. Gynecol. **20**(5), 452–457 (2002)
47. Bamfo, J.E.A.K., Kametas, N.A., Turan, O., Khaw, A., Nicolaides, K.H.: Maternal cardiac function in fetal growth restriction. BJOG Int. J. Obstet. Gynaecol. **113**(7), 784–791 (2006)
48. Melchiorre, K., Sutherland, G.R., Liberati, M., Thilaganathan, B.: Maternal cardiovascular impairment in pregnancies complicated by severe fetal growth restriction. Hypertension. **60**(2), 437–443 (2012)
49. Nagueh, S.F., Appleton, C.P., Gillebert, T.C., et al.: Recommendations for the evaluation of left ventricular diastolic function by echocardiography. J. Am. Soc. Echocardiogr. **22**(2), 107–133 (2009)
50. Krause, K., Möllers, M., Hammer, K., et al.: Quantification of mechanical dyssynchrony in growth restricted fetuses and normal controls using speckle tracking echocardiography (STE). J. Perinat. Med. **45**(7), 821–827 (2017)
51. Crispi, F., Figueras, F., Cruz-Lemini, M., et al.: Cardiovascular programming in children born small for gestational age and relationship with prenatal signs of severity. Am. J. Obstet. Gynecol. **207**, 121.e1–121.e9 (2012)
52. Cikes, M., Sutherland, G.R., Anderson, L.J., Bijnens, B.H.: The role of echocardiographic deformation imaging in hypertrophic myopathies. Nat. Rev. Cardiol. **7**, 384–396 (2010)

53. Barker, D.J.: The developmental origins of well-being. Philos. Trans. R. Soc. Lond. Ser. B Biol. Sci. **359**, 1359–1366 (2004)
54. Cosmi, E., Visentin, S., Fanelli, T., Mautone, A.J., Zanardo, V.: Aortic intima media thickness in fetuses and children with intrauterine growth restriction. Obstet. Gynecol. **114**, 1109–1114 (2009)
55. Jarvisalo, M.J., Jartti, L., Nanto-Salonen, K., et al.: Increased aortic intima-media thickness: a marker of preclinical atherosclerosis in high-risk children. Circulation. **104**, 2943–2947 (2001)
56. Comas, M., Crispi, F., Cruz-Martinez, R., Figueras, F., Gratacos, E.: Tissue Doppler echocardiographic markers of cardiac dysfunction in small-for-gestational age fetuses. Am. J. Obstet. Gynecol. **205**, 57.e1–57.e6 (2011)
57. Nilsson, P.M., Ostergren, P.O., Nyberg, P., Söderström, M., Allebeck, P.: Low birth weight is associated with elevated systolic blood pressure in adolescence: a prospective study of a birth cohort of 149,378 Swedish boys. J. Hypertens. **15**, 1627–1631 (1997)

Chapter 4
Cardiotocography for Fetal Monitoring: Technical and Methodological Aspects

Giovanni Magenes and Maria G. Signorini

Contents

4.1 Introduction

The main goal of any kind of fetal monitoring is to assess fetus well-being condition in order to minimize risks of fetal morbidity and mortality, to identify "at-risk" fetuses, and to evaluate the optimal time of delivery. Among the indirect information that can be detected during pregnancy and labor, the fetal heart rate (FHR) signal is certainly one of the most reliable sources on the health status of the fetus. Various measurement techniques have been used to detect the FHR (indirect electrocardiographic, ultrasonocardiographic, phonocardiographic); however, the

G. Magenes (✉)
Department of Electrical, Computer and Biomedical Engineering, University of Pavia, Pavia, Italy
e-mail: giovanni.magenes@unipv.it

M. G. Signorini
Department of Electronics, Information and Bioengineering (DEIB), Politecnico Milano, Milan, Italy
e-mail: mariagabriella.signorini@polimi.it

© Springer Nature Switzerland AG 2021
D. Pani et al. (eds.), *Innovative Technologies and Signal Processing in Perinatal Medicine*, https://doi.org/10.1007/978-3-030-54403-4_4

ultrasound technique with a Doppler probe is the one that has found wider use in clinical practice both for its characteristics of easy usability and for its wide range of applicability.

Why is it fundamental to check FHR to assess fetal well-being? It is known that the fetal brain modulates the FHR through an interplay of sympathetic and parasympathetic actions of the autonomous nervous system. Thus, FHR monitoring can be used to determine if the fetal brain is well oxygenated. In case of fetal distress during labor, the FHR shows some morphological anomalies. If FHR does not show anomalies, chances are high that the fetus can stand the labor.

Moreover, several conditions such as hypoxia, acidemia, and drug induction produce noticeable variations of FHR, which are visible even by simple eye inspection. As a general consideration, we can affirm that fetal distress is preceded by alterations in the RR intervals time series before any pathological change in fetal condition definitely occurs.

The cardiotocography (CTG) is nowadays the most used technique in the developed countries to monitor fetal condition through the measurement of FHR during the antepartum period and during the labor (more than 90% of pregnant women are monitored at least once during the pregnancy and almost all are monitored during labor). The CTG consists of the simultaneous recording of fetal heart rate (FHR) and uterine contractions (tocogram) in order to check the FHR variations in conjunction with the forces exerted by the contractions, which can modify the FHR.

Since the introduction in the clinical practice of the first commercial cardiotocograph in Europe, thanks to Hammacher in 1968 [1], a great scientific commitment and a lot of enthusiasm have been lavished on this technology, which seemed to give a concrete possibility of "understanding" when and why it occurred a deterioration of fetal health, and therefore to intervene more effectively. Cardiotocographic monitoring (CTG), or cardiotocography, starting from the '70s, has had a considerable diffusion in clinical practice, so that it remains, to date, the most widely used method for checking fetal condition over time, despite the advent of other more recent techniques.

Various studies have shown the validity of CTG in the first period of labor during delivery in predicting fetal hypoxia. Taking into account the large amount of monitored childbirths, it can be reasonably concluded that a "normal" CTG pattern is a good indicator of fetal well-being, while a pattern showing a low FHR variability can indicate a fetal sufferance, although only in 50% of cases [2].

On the other hand, the effectiveness of antepartum cardiotocography, or non-stress test, in identifying fetal distress during pregnancy presents characteristics that are more controversial. The main limitations are not so much due to the technique itself, but to the difficulties in reading and interpreting the FHR signal, whose variations originate from the interaction of numerous and complex physiological mechanisms. The intrinsic complexity of the FHR signal, combined with a qualitative analysis, inevitably led to considerable difficulties in formulating homogeneous judgments among different clinicians and to scarce results in terms of diagnostic forecasting. There are numerous physiopathological and technical factors, which, if not correctly taken into consideration in the evaluation of the CTG trace, lead to inconsistencies between the result of a CTG test and the neonatal outcome, with consequent incorrect patient management. The major problem is the lack of general agreement regarding the reading and interpretation criteria of CTG tracings. In fact, since the introduction of the CTG technique in clinical practice, a lot of visual reading methodologies have been proposed, none of which has clearly established itself as a clinical "golden" standard, despite the indications given by the most diffused obstetrical associations (ACOG, FIGO, etc.). Indeed, several studies have shown that, even in centers that adopt the same rules, individual observers rarely agree on the evaluation of the same tracing [3, 4]. Furthermore, there is great evidence that the visual examination of the CTG trace is not able to extract all the cardiac variability information contained in the FHR signal. Features such as the magnitude of the periodic components of the signal generated by the cardiac pacemaker, the nonlinearity of the FHR control system, or even the short-term variability by themselves cannot be grasped by a simple eye inspection of the CTG tracing.

Although cardiotocography (CTG) has been used since the 1970s as a noninvasive method to monitor the status of the fetus and has allowed a drastic reduction in early intrapartum and neonatal mortality, its diagnostic accuracy is still far from being fully satisfactory [5].

In response to these problems, the use of numerical systems for automated analysis of fetal heart rate was introduced first as an experimental method and currently also in clinical practice. Numerous centers in the world have developed and proposed different computerized systems, some of which have also reached commercialization and have been extensively tested at the level of multicentric research.

In this chapter we will illustrate the basic principles of the computerized CTG technique and the methodologies of FHR analysis, taking into account our experience acquired in almost 20 years of research.

4.2 Computerized Cardiotocography

A system for numerical (or computerized) cardiotocography is logically and structurally composed of two devices, the cardiotocograph, by which the FCF and tocodynamometry (toco) signals are detected, and a computer by which the signals are analyzed. The two devices can be physically separated in different containers, as represented in Fig. 4.1 or they can be industrially assembled in a single case.

In CTG systems, the FHR is derived from the time series of the beat-beat intervals identified by the cardiotocograph and computed in milliseconds. Its value in beats per minute (bpm) is calculated by the inversion of the interbeat value T according to the formula FHR (bpm) = 60,000/T(ms). The frequency at which the cardiotocograph updates the FHR values differs from system to system and varies from 10 Hz to about 0.2 Hz. The most diffused cardiotocographs, derived from the Hewlett Packard series 135X (Agilent, Philips, General Electric-Corometrics, etc.), adopt the same protocol and give an FHR value each 250 ms (4 Hz). This update time depends on the firmware installed on the cardiotocograph. As a matter of fact,

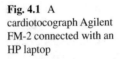

Fig. 4.1 A cardiotocograph Agilent FM-2 connected with an HP laptop

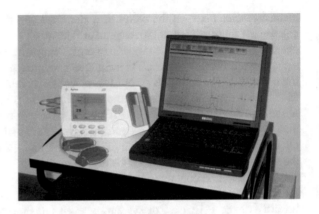

in Doppler detection systems, the recognition of the interbeat interval is carried out through the autocorrelation of the received echo signal, which is affected by variations also due to small movements of the fetus or mother. The use of the autocorrelation function represents a compromise between accuracy in identifying the beat-beat interval and measurement reliability. It was introduced in order to improve the signal/noise ratio when the quality of the return echo is poor and to eliminate the artifacts caused by the different moving structures of the fetal heart. The autocorrelation allows to identify a single peak within the Doppler signal of a heartbeat and therefore does not present ambiguity in the determination of the beat-beat interval. In case of low signal/noise ratio, the autocorrelation function allows to extrapolate the temporal position of the peak to be used for the calculation of the FHR considering the beat-beat intervals previously recorded and superimposing them on the poor quality signal that is received at that moment. This extrapolation allows in most cases to still provide an FHR value close to the true, but obviously non-existent as obtained from the average of previous values.

Therefore, the FHR values obtained with the autocorrelation are unable to follow the sudden changes in the FHR (low-pass filtering effect of cardiac variability).

In order to give technical data about the Doppler probe and the autocorrelation procedure, we report here the specifications of the HP -series 135X cardiotocographs:

Hardware – The US transducer transmits 998.4 kHz ultrasound bursts. The burst widths are controlled by firmware. The repetition rate is 3.2 kHz. The received echo signal is amplified by a high-frequency amplifier with a gain of 120. The demodulator is controlled by software in its receive window. The demodulated LF signal is bandpass filtered (100–500 Hz) and amplified by a software-controlled gain.

Software – The Series 50 fetal monitors (M1350A/B/C, M1351A, M1353A) use an autocorrelation technique to compare the complete ultrasound Doppler signal of a heartbeat with the next one. A peak detection software then determines the heart period from the autocorrelation function. The Doppler signal is sampled

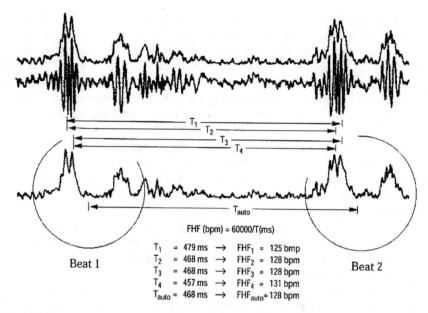

FHF (bpm) = 60000/T(ms)

T_1	= 479 ms	→	FHF_1 = 125 bmp
T_2	= 468 ms	→	FHF_2 = 128 bpm
T_3	= 468 ms	→	FHF_3 = 128 bpm
T_4	= 457 ms	→	FHF_4 = 131 bpm
T_{auto}	= 468 ms	→	FHF_{auto} = 128 bpm

Fig. 4.2 Comparison of beat-to-beat intervals computed with peak detection techniques and with autocorrelation. The Envelope Doppler signal (upper trace) was derived from the demodulated Doppler signal (second trace) In the lower trace, four possible heart periods T1–T4 can be identified by peak detection. Symbolized by the two circles the heart rate interval is correctly calculated by autocorrelation of the waveform complexes

with 200 Hz (5 ms). With a peak position interpolation algorithm, the effective resolution is better than 2 ms.

In Fig. 4.2, an example of the identification of beat-beat interval through autocorrelation on the demodulated Doppler signal is shown.

The detection of the activity of the autocorrelation function at each FHR value provided by the cardiotocograph is therefore very useful for the clinician to establish how real or reconstructed the signal is. In some systems the signal actually measured is highlighted on the screen in green, while the one reconstructed in yellow.

A parameter closely related to the previous one, and equally important, is the quality of the recording, expressed numerically as a percentage of "lost", or unregistered, beats on the entire track. It is known that the quality of CTG registration by Doppler is in inverse correlation with the gestational age: in fetuses of gestational age of less than 30 weeks, i.e., those who in case of growth retardation or other pathologies will mostly meet perinatal outcomes, the percentage of signal loss can reach 20% of the total, thus invalidating the reliability of the method. The loss of signal (when even the autocorrelation function cannot obtain an extrapolated FHR value) is normally highlighted on the screen in red.

Commercially available computerized systems generally use relatively low updating rates, up to 3.75 seconds of the Sonicaid System 8000–8002. In general, it is not necessary to increase the updating frequency, because the Doppler detection of the FHR signal by means of an autocorrelation function entails, as already mentioned, a limitation in the frequency band of the beat-beat variability [6]. However, from comparisons made experimentally among different cardiotocographs (HP, Sonicaid, Corometrics, Nihon Koden, Wakeling), it has been noted that the accuracy in chasing the variation of beat-beat intervals is very different in the various models of cardiotocograph [7] and, while in some cases it would not have meant increasing the sampling frequency, in others it is also possible to reach 4 Hz.

In essence, using a cardiotocograph with FHR Doppler measurement, it is not useful to sample the FHR at intervals lower than the amplitude of the time interval on which the autocorrelation function acts, which is carried out differently in the various cardiotocographs; this time can vary from 250 msec to more than 1 second depending on the model [8].

In our system (2CTG2 system) or experimental systems, a useful band of at least 1 Hz is considered so as to allow a fairly precise assessment of the "real" beat-to-beat variability and the use of advanced analysis techniques such as spectral analysis or the study of FHR regularity [9].

The sampling of the signal relating to the uterine tone ("toco" channel) is normally carried out at the same frequency as the FHR signal in order to have a correct alignment of the data (for each FHR value there is a corresponding toco value), even if the band of toco signal frequencies is lower than that of cardiac variability.

It is also possible to monitor fetal motor activity, displaying fetal movements on the CTG tracing and allowing their visual and automated correlation with FHR events (accelerations, decelerations, variability). Motor activity can be recorded with a subjective detection method, by making the mother operate a button every time she hears the fetus move, or automatically by the cardiotocograph through ultrasound detection and use of different mathematical algorithms for fetal movement identification [10].

Most cardiotocographs transmit through a serial line (RS-232, RS-485, USB, etc.) the following digital information to the computer, with an asynchronous protocol (usually the standard is Hewlett Packard protocol):

- FHR values expressed in bpm
- Toco values expressed in arbitrary units
- Quality of the signal (defined as a number)
- Presence of fetal movements as time marks.

In the HP protocol, all values are represented by text ASCII characters.

The computer must be equipped with a software program capable of reading the FHR and toco digital signals, making changes and calculations on them (eliminating noise or artifacts, checking the quality of the signal, etc.), mathematically analyzing the tracings by extracting the significant parameters and storing the data on magnetic media, generally in the form of "records" of a database also containing the main registry and clinical information relating to the patient under study. The newly developed software allows a quasi-real–time visualization of the cardiotocographic trace and the analysis results directly on the screen, updating the calculations while the CTG ("real–time monitoring") recording is in progress. Since the track is stored on the computer, at the end of the recording it is possible to re-analyze it also by going to select portions of the same, chosen by the user.

Finally, it should be emphasized that the latest generation computers are potentially capable of acquiring multiple CTG exams at the same time, i.e., it is possible to connect several cardiotocographs to the same computer at the same time. Obviously, the software must provide for the management of several contemporary exams.

Both FHR and toco signals are submitted to various algorithms in order to extract the relevant morphological events and/or the parameters of FHR, specific of each CTG system. Unfortunately, there is no agreement on a unique methodology to extract this information, because most morphological features are identified by trying to reproduce the "eye-inspection" criteria reported by the Obstetrical Associations, which are by definition based on a qualitative interpretation of the tracings. In the following paragraphs, we illustrate how our CTG system computes a set of morphological features (baseline, accelerations, and decelerations) and parameters related to the time domain characteristics of the FHR, with those in the frequency domain and with the nonlinear nature of the heart rate variability.

The software was developed in collaboration with Hewlett Packard Italy starting at the beginning of the year 2000 [11] and it is now used for research purposes on antepartum CTG in our group and in several Italian clinical centers.

4.3   FHR Morphological Analysis

4.3.1   Baseline FHR and Baseline Measurements

The NICHD (National Institute of Child Health and Human Development) nomenclature [12] defines baseline fetal heart rate as: *"the baseline FHR is determined by approximating the mean FHR rounded to increments of 5 beats per minute (bpm) during a 10-minute window, excluding accelerations and decelerations and periods of marked FHR variability (greater than 25 bpm). There must be at least 2 minutes of identifiable baseline segments (not necessarily contiguous) in any 10-minute window, or the baseline for that period is indeterminate. In such cases, it may be necessary to refer to the previous 10-minute window for determination of the baseline. Abnormal baseline is termed bradycardia when the baseline FHR is less than 110 bpm; it is termed tachycardia when the baseline FHR is greater than 160 bpm"*.

In practice, the baseline identifies a hypothetical average (sinus) trend of the FHR signal, once it has been purified of events capable of altering its behavior.

In the design of a computerized CTG system, a reproducible and reliable determination of the baseline is a fundamental starting point, because it influences the correct identification of accelerations and decelerations. Several attempts in this direction have been made starting from the work of Dawes et al. [13]. In our 2CTG2 system we followed the algorithm suggested by Mantel et al. [14], but we tuned the parameters of the algorithm, to make the outcome fully compliant also with the opinions of our team of clinicians, expert on CTG analysis. The algorithm is very complex, and a full description can be found in the cited reference. The result of this analysis is a numerical signal, indicated as the baseline FHR (expressed in beats/ minute, bpm) starting from which the accelerations and decelerations are identified.

In particular, any overestimation of the baseline is reflected in a significant underestimation of the number of accelerations while the number of decelerations will be overestimated. This error certainly affects the visual evaluation but also, in part, the analysis carried out by the computerized systems and is more frequently found in traces that are characteristic of the fetus in the active waking phase. The consequence is a possible excess of invasive assessments until the unjustified completion of the birth (false positive).

The dual error, i.e., that of underestimation, occurs more rarely. In this case, there would be an estimate by default of the number of decelerations, with a consequent lack of necessary diagnostic and/or therapeutic intervention (false negative) with possible serious consequences.

In recent systems, the baseline is estimated incrementally, as registration proceeds, by means of a multi-pass digital filter that can be implemented in various configurations.

4.3.2   Accelerations and Decelerations

Accelerations are events on the CTG track during which the FHR expressed in bpm remains persistently above the estimated baseline value, for a prolonged period (in seconds). Both the minimum amplitude, above which the FHR must remain, and the residence time are variable parameters according to the system used.

The NICHD definition [12] of acceleration is: "*a visually apparent abrupt increase in fetal heart rate. An abrupt increase is defined as an increase from the onset of acceleration to the peak in less than or equal to 30 seconds. To be called an acceleration, the peak must be greater than or equal to 15 bpm, and the acceleration must last greater than or equal to 15 seconds from the onset to return to baseline* [15]. A prolonged acceleration is greater than or equal to 2 minutes but less than 10 minutes in duration. An acceleration lasting greater than or equal to 10 minutes is defined as a baseline change. Before 32 weeks of gestation, accelerations are defined as having a peak greater than or equal to 10 bpm and a duration of greater than or equal to 10 seconds.

A further subdivision of the accelerations is obtained by establishing different threshold criteria to quantify the increase in the FHR in bpm with respect to the estimated value for the baseline: a subdivision into small and large accelerations is thus obtained. For example, an increase in the FHR above the baseline greater than 15 bpm for at least 15 seconds is the definition of great acceleration according to

Arduini et al. [16]. Depending on the threshold value and the selected time window, the different systems will produce a different quantification of the events.

In a completely analogous way, the identification of the decelerations is obtained.

CTG systems usually provide the number of accelerations and decelerations, normalized to 60 minutes of tracing, and for each event, the duration, the maximum distance from the baseline, the area of acceleration or deceleration expressed in beats per minute, and, as regards decelerations, the distance of the nadir from the peak of the previous contraction (lag time).

Normally the computerized system, before identifying an acceleration or deceleration as real, verifies that the quality control of the recorded signal, previously described, has given a positive result [17].

4.4 Analysis of FHR Variability

The indices just described are basically the reproduction of the criteria used for visual analysis, the implementation of which has not introduced elements of substantial novelty except the decrease of the analysis time of the tracings and the removal of interobserver variability.

Both the importance and novelty of automatic computerized analysis lie in the significant improvement in extracting the information content from the FHR variability signal. In fact, the duration of the heartbeat varies physiologically over time through the action of complex mechanisms both spontaneous and activated by different stimuli. As depicted in Fig. 4.3, a lot of quantitative indices can be computed on the FHR signal, which contains information on the nervous control mechanisms that generated it, determining its characteristics both in the short and long term.

As observed in adults, the sympathetic branch of the autonomic nervous system acts by inducing an increase in heart rhythm through vasomotor control and baroreceptive reflex [17]. The action of the parasympathetic branch (which for the heart coincides with the vagus nerve) regulates the slowing down of the heart rate. The two systems work synergistically to ensure cardiac response in different physiological conditions and at different time scales. Observations on animals and humans in borderline conditions (e.g., after cardiac transplantation [18] have shown that, after the suspension of vagal and sympathetic efferences, the heart beats in constant sinus rhythm, with variability heart rate almost absent.

In the fetus, the nervous system has yet to complete the development phase. Nevertheless, the variability of the FHR, or the sequence of estimated time duration values between two successive heartbeats, is an indication, in a healthy fetus, of the action of regulation of the sympathetic and parasympathetic nervous system on the sinus node.

The ability and rapidity of variation of the heart rate are related to the adaptability of the fetal heart and are therefore of great clinical importance.

In addition to the synergistic activity of the autonomic sympathetic and vagal nervous system, various physiological factors have an influence on the variability of

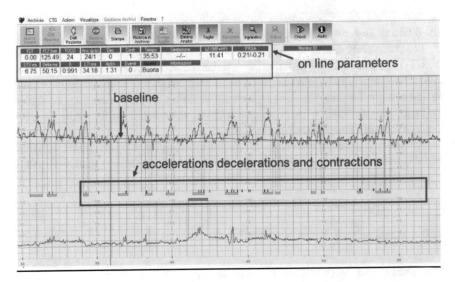

Fig. 4.3 Screenshot of a CTG exam from the 2CTG2 system. The FHR baseline (in blue) is drawn by means of the Mantel's algorithm; small black arrows indicate accelerations. The time duration of accelerations and contractions is depicted respectively with green and blue rectangles. The various CTG parameters are reported in the bar below the Menu of the software

FHR. The amplitude of the variability is therefore conditioned by the interaction between respiratory influences (mediated by the action of the vagus nerve) and behavioral and motor influences in which the control mechanism is mainly that of the sympathetic system. The activity of fetal respiratory movements (when present) modulates the heart rate signal and modifies its variability [19]. Another mechanism is the maternal breath which influences the fetal heart rate so that frequency components identify frequency components fetal heart rate related to maternal respiratory rate [20].

Fetal body movements also introduce components into the FHR signal. Activities such as movements of the mouth (attempts to suck), limbs, and facial grimaces are all correlated with an increase in the sympathetic contribution to the variability of FHR.

Variability changes with gestational age, both in terms of the temporal duration of the phases and in the number of accelerations [21].

Furthermore, FHR variability increases as a result of vibroacoustic stimulation of the fetus with laryngophone on the maternal abdomen [22]. This behavior is also documented in the literature for other types of sensory stimulation [23].

Pathological conditions, such as hypoxia and anoxia, taking drugs with central sedative action, the immaturity of the hypothalamic and medullary centers, and brain malformations, can affect the vasomotor center and the medullary nuclei of the vagus with a consequent alteration (reduction) of FHR variability.

The analysis of such complex interactions can be carried out only by means of the tools made available by the computerized analysis.

Heart rate variability is a phenomenon that can be assessed at different time scales. We therefore speak of short, medium, and long-term variability according to the number of samples of the heart rate signal which is necessary for measuring the parameter.

The measurement of the difference between successive FHR values is the basis for calculating variability indicators in the short term (over a few beats or for a few seconds) whose value can be expressed in beats per minute (bpm), milliseconds (msec) or using dimensionless indices.

4.4.1   Classical Time Domain Variability Indices

In many CTG systems, each FHR value coming from the cardiotocograph is transformed in equivalent RR interval and expressed in milliseconds for the computation of time domain parameters. In the following, we will make reference to the algorithms adopted by our 2CTG2 system for computing the FHR variability indices.

Short-Term Variability (STV) [ms]: it quantifies FHR variability on a short time scale. Considering an interbeat sequence of 1-minute duration, STV is defined as:

$$STV = \text{mean}\left[\left|T(i+1)-T(i)\right|\right]_i = \frac{\sum_{i=1}^{23}\left|T(i+1)-T(i)\right|}{23}$$

where T is the average FHR computed by dividing the FHR recording in non-overlapping windows of 5 consecutive FHR values (2.5 s for a sampling frequency of 2 Hz). *STV* is computed in a window of 1-minute duration so that 24 T values are obtained for each window. The corresponding STV estimate is obtained by averaging the differences between adjacent $T(i)$ values, having accelerations and deceleration excluded.

Interval Index (II): it provides an estimate of short-term variability scaled by STV, defined as:

$$II = \frac{\text{std}\left[\left|T(i+1)-T(i)\right|\right]_i}{STV}$$

where $i = 1, ..., 23$ are the total number of T FHR values recorded in 1 minute.

Delta [ms]: considering a window in time of 1-minute duration, Delta is defined as:

$$\text{Delta} = \max\left(T(i)\right) - \min\left(T(i)\right)$$

where $i = 1, ..., 24$ which are the total number of T FHR values recorded in 1 minute.

Long-Term Irregularity (LTI) [ms]: is defined as the interquartile range of the distribution $m(j)$ which is defined as:

$$m(j) = \sqrt{T^2(j+1) + T^2(j)}$$

where $j = 1, \ldots, 71$ which are the total number of T FHR values in a 3-minute window. LTI quantifies FHR variability on a longer time scale with respect to the previously reported time domain indices excluding accelerations and decelerations.

All these time domain parameters are computed following the indications of Arduini et al. [16].

4.5 The Need for a Novel Approach

As already described in the previous paragraph, time domain variability indices are the history of the FHR analysis. Short-time variability (STV), long-term irregularity (LTI), Delta, and interval index (II) represent a quantification of the qualitative reading by eye inspection.

Furthermore, heart rate analysis in adult subjects, both healthy and affected by diseases, has taken a different path with important findings about the possibility to quantify autonomic nervous system mechanisms and their impairment in disease states.

As the development of the ANS is in progress even in the fetal growth, the extension to FHR of the analysis methods, that already demonstrated usefulness in adults, opened new perspectives on the fetal diagnosis and the assessment of fetal well-being.

These novel parameters should consider FHR time series as complex signals in which different, time, frequency, and even nonlinear mechanisms contribute. In fact, all these contributions produce effects that can be captured by methods and parameters extracted in different domains.

The point is to go beyond the traditional time domain by adding features in frequency, nonlinear, and complexity sphere, which are not available without advanced computational capabilities.

This is a new route to classify the FHR signal that takes advantage of novel signal processing methods, whose features can be associated with specific pathophysiologic events in the heart control mechanisms.

Our work during the last 20 years was devoted to the applications of this novel set of indices to the analysis of the FHR signal.

4.5.1 FHR Features in Frequency Domain

Other HR variability parameters are calculated in the frequency domain following the Task Force indications by Power Spectral Density (PSD) [17]. PSD can be estimated using the Fourier transform (usually through FFT algorithm) obtaining non-parametric estimations based on the direct estimation of the Periodogram in the FHR. Parametric methods instead estimate PSD models from FHR and evaluate the related parameters. Autoregressive power spectrum estimation belongs to this second group.

Some of them in specific frequencies show a relationship with ANS components whose activity can be quantified through frequency and power (contribution in the signal variance) values. For the HRV signal, main variability rhythms are LF (0.1 Hz – cardiovascular control) and HF (0.25 Hz –breathing).

In the context of FHR analysis, modification of standard analysis bands for HRV is required. Thus frequency domain parameters are the power percentage of the following frequency band ranges: LF [ms^2/Hz] in (0.04–0.15 Hz), connected with sympathetic nervous system; MF (movement frequency [ms^2/Hz]) in (0.15–0.5 Hz, not present in adult human subjects) which is associated with maternal breathing and fetal movements; HF power [ms^2/Hz] in (0.5–1.0 Hz) linked to parasympathetic nervous system and fetal respiration-like activity. LF/(MF + HF) ratio provides the estimation of sympathovagal balance [9].

Each FHR value coming from the fetal monitor (in our case HP series 1351A) was transformed in equivalent RR interval and expressed in milliseconds for the computation of frequency domain parameters.

PSD is a widely employed tool for HRV frequency analysis as it can quantitatively measure the periodic oscillations related to neural control activity, namely autonomic nervous system (ANS) modulation over the cardiac system.

In the context of this analysis, the PSD estimation for FHR was performed based on autoregressive (AR) modeling (parametric spectral estimation). The AR model utilized to mimic FHR dynamic is defined as:

$$\hat{FHR}(j) = \sum_{p}^{i=1} a_i \cdot FHR(n-i) + w_j$$

where $w_j \sim$ WGN(0, σ^2) (white Gaussian noise), p is the model order (from 8 to 12), and a_i are the model parameters. The modeled FHR windows ($\hat{FHR}(j)$) are of duration equal to 3 minutes ($j = 1, ..., 360$).

Model parameters are calculated recursively by means of the Levinson-Durbin algorithm. Once the proper model order is defined, so that the model parameters are determined, PSD is defined as:

$$PSD(f) = \frac{\sigma^2 \Delta}{\left|1 - \sum_{k=1}^{p} a_k e^{-j2\pi kf\Delta}\right|^2} = \frac{\sigma^2 \Delta}{A\left(e^{j2\pi f\Delta}\right) A^*\left(e^{j2\pi f\Delta}\right)} = \frac{\sigma^2 \Delta}{A(z) A^*\left(\frac{1}{z^*}\right)}\Bigg|_{z=e^{j2\pi f\Delta}}$$

where Δ is the mean value of $\hat{FHR}(j)$ in seconds and $A(z)$ is the z-transform of the transfer function of the AR process previously defined. Through this parametric approach, FHR signal undergoes an automatic decomposition into a sum of sinusoidal contributions, themselves identified by their corresponding central frequencies and the associated power [24].

AR model estimation of HR interval series decomposes the power in harmonic components each one characterized by its central frequency and an associated amount of power. The order of the model (p) identifies the number of harmonic components that are necessary to explain the power (variance) content of the signal in the frequency domain.

4.5.2 Nonlinear Domain Parameters

Not only frequency and time domain parameters explain the variety of behaviors that can be observed in FHR, it is now assessed that nonlinear mechanisms are involved in the variability generation.

For this reason, it is useful to add features derived from nonlinear analysis methods. Among them, measure of complexity rate in the FHR signal and the evaluation of periodic unstable oscillations provide powerful parameters improving discrimination and classification ability. In the context of a novel monitoring approach, these features play an important role.

Entropy estimators measure the signal regularity. Regularity has some correlations with complexity of the system generating the signal. An irregularity increase characterizes a shifting of the system toward a random behavior. Irregularity decrease indicates a loss of system complexity (presence of pathological reduction of the variability, i.e., sudden cardiac death, gait failure, elderly condition, etc.).

The entropy estimators adopted in this work are the approximate entropy [25] and the sample entropy [26].

Entropy indices can be calculated both on very short periods (few minutes, at least 300 samples) and longer intervals. These statistics depend on parameters r and m: m is the detail level at which the signal is analyzed, and r is a threshold, which filters out irregularities. The adopted values are $m = 2$ and $r = 0.15*SD$, where SD is the standard deviation of the original signal.

The multiscale entropy (MSE) captures the fluctuations at different degrees of resolution, i.e., in a multiscale manner [27], allowing the study of the signals structure. The method creates coarse-grained time series $\{y\tau\}$. The length of these series is the ratio between the length of the original series and τ. The coarse-grained time series are constructed by averaging consecutive τ samples. For each coarse-grained time series, an entropy measure is estimated.

Each FHR value coming from the CTG monitor (HP –series 1351A) was transformed in equivalent RR interval and expressed in milliseconds for the computation of approximate entropy and Lempel and Ziv complexity.

Approximate Entropy (ApEn) [bits]: quantifies a signal regularity by assessing the occurrence rate of patterns by comparing the patterns themselves to a reference one of length m. Pattern similarity is defined based on a threshold r [4].

Given a sequence of N data points $u(i)$, $i = 1, ..., N$, the algorithm creates sequences $x_m(i)$ (based on window length m) and it computes for each $i \leq N - m + 1$ the quantity expressed as:

$$C_i^m(r) = \frac{1}{N}\left\{\text{count of } j \leq N - m + 1 \,|\, \text{distance}\left[x_m(i), x_m(j)\right] \leq r\right\}$$

Approximate entropy (ApEn) is defined as:

$$\text{ApEn}(m,r) = \lim_{N\to\infty}\left[\Phi^m(r) - \Phi^{m+1}(r)\right]$$

where $\Phi^m(r) = \dfrac{1}{N-m+1}\sum_{N-m+1}^{i=1}\ln\left(C_i^m(r)\right)$.

In the context of finite time series of length N as for FHR, ApEn can be written as:

$$\text{ApEn}(m,r,N) = \Phi^m(r) - \Phi^{m+1}(r)$$

In this work, ApEn was estimated by considering non-overlapping windows of duration equal to 3 minutes, with the following parameter setting: $m = 1$, $r = 0.1$, $N = 360$ and named *ApEn(1,0.1)* [28, 29].

Lempel and Ziv complexity (LZC) [bits]: it quantifies the rate of new patterns developing with the evolving of the signal [30]. LZC is a method originally proposed to assess the algorithmic complexity. Its value increases as the gradual increase of new patterns along the sequence. The algorithmic complexity is defined by the information theory as the minimum quantity of information needed to define a binary string. For example, the algorithmic complexity of a random string is the length of the string itself. In this case, any compression effort will produce an information loss.

The first step toward its formulation encompasses the definition of the quantity $c(n)$ which measures the number of different substrings and the rate of their recurrence in a given time series. According to the information theory, in turn it assesses the minimum quantity of information needed to define a binary string. LZC quantifies the rate of new patterns arising as signal evolves [30].

Suppose the number of symbols in the alphabet A is α and the length of sequence is equal n. The upper bound for $c(n)$ is given by:

$$c(n) < \frac{n}{(1 - \varepsilon_n)\cdot\log(n)}$$

where

$$\varepsilon_n = 2\frac{1+\log\left(\log\left(\alpha n\right)\right)}{\log\left(n\right)}$$

and log(x) means the logarithm of x to the base α.

When n is large enough ($n \rightarrow \infty$), $\varepsilon_n \rightarrow 0$ so as a result:

$$b\left(n\right) = \lim_{n\to\infty} c\left(n\right) = \frac{n}{\log\left(n\right)}$$

where $b(n)$ is the asymptotic behavior of $c(n)$ for a random string.

The normalized complexity is thus defined as:

$$C\left(n\right) = \frac{c\left(n\right)}{b\left(n\right)}$$

In order to obtain LZC estimation for FHR time series, the latter requires to be transformed into a symbolic sequence according to a binary and/or a ternary coding procedure.

Binary coding: given an FHR series $x(N)$, the sequence $y(N)$ is built by assigning 1 when the condition $x(n + 1) > x(n)$ is verified for $n = 1, \dots, N$. On the opposite case of signal decrease, $y(n)$ is assigned to 0 when the condition $x(n + 1) \leq x(n)$ is met.

Ternary coding: given an FHR series $x(N)$, the sequence $y(N)$ is built as in the binary coding case with the additional condition of signal invariance which is defined as $x(n + 1) = x(n)$ and coded with the symbol 2.

Additionally, in the context of recorded time series, a factor p is introduced to define the minimum quantization level for a symbol change in the coded string (e.g. $y(n) = 1$ if $x(n + 1) > x(n) + x(n) \cdot p$).

In this work, LZC was estimated by considering non-overlapping windows of duration equal to 3 minutes, with the following parameter setting: binary coding and $p = 0$ and named LZC(2,0). The choice of $p = 0$ reflects the current value for the quantization level, which is actually ±0.5 bpm [31].

4.5.3 Phase-Rectified Signal Averaging

PRSA quantifies quasi–periodic oscillations in non-stationary signals affected by noise and artifacts, by synchronizing the phase of all periodic components [32].

It requires a time series $I = 1,\dots,N$, characterized by periodicities and correlations, containing non-stationary and noise events. The first step is the computation of anchor points (AP) selected based on the average value of the signal before and after a certain instant k within a selected time window. AP are identified as points that mark a signal increase. Similar inequality for signal decrease (< symbol). In an experimental time series as the HRV, about half of the signal points are anchor points. Features are denoted as acceleration and deceleration phase-rectified slope (APRS and DPRS) such as the slope of the PRSA curve in its AP *(1)*.

We define the acceleration (deceleration) phase-rectified slope (APRS or DPRS) as the slope of the PRSA curve computed in the anchor point. The two parameters describe both the average increase (decrease) in FHR amplitude (absolute change of heart frequency) and the time length of the increase (decrease) event.

Each FHR value coming from Hewlett Packard CTG fetal monitors (series 1351A) was expressed in bpm for the computation of acceleration phase-rectified slope and deceleration phase-rectified slope in order to be concordant with the common definition of acceleration and deceleration in fetal heart rate monitoring.

Acceleration Phase-Rectified Slope (APRS) [bpm]: the computation of phase-rectified signal averaging (PRSA) curve (which APRS is extracted from) starts from considering a time series x_i of length N $(i = 1, ..., N)$ as FHR in this work. The first step toward PRSA computation is the determination of the so-called anchor point (aPs). In this context, aPs are defined as the time series points x_i fulfilling the following inequality:

$$\frac{1}{M} \sum_{M-1}^{j=0} x_{i+j} > \frac{1}{M} \sum_{M}^{j=1} x_{i-j}$$

where the parameter M is employed as the upper frequency bound for the periodicities to be detected by PRSA method.

After aPs being detected, windows of length $2L$ are built symmetrically with respect to each AP. Given the fact that the majority of aPs are temporally close one each other, the resulting windows are effectively overlapping. An additional specification for the parameter L is that it should be larger than the period of slowest oscillation to be detected [33].

The PRSA curve X_i is obtained by averaging the derived windows synchronized in their aPs. After obtaining the PRSA curve, it is useful to summarize its characteristics by extracting different parameters. An example of such is APRS defined as:

$$APRS = \frac{\partial X_i}{\partial i} |_{i=AP}$$

The parameter APRS is a descriptor of the average increase in FHR amplitude and the time span of such increased event.

In this work, the considered signals x_i is the whole available FHR recording, thus resulting in a single APRS value. The parameters M and L are equal 40 and 200 respectively [34].

Deceleration Phase-Rectified Slope (DPRS) [bpm]: the computations are analogous of those previously reported for APRS apart from the definition of aPs which are defined as the time series points x_i fulfilling [33]:

$$\frac{1}{M}\sum_{M-1}^{j=0} x_{i+j} < \frac{1}{M}\sum_{M}^{j=1} x_{i-j}$$

The use of these advanced methods allows collecting a set of features that analyze different properties of the FHR variability signals.

Table 4.1: Indices applied in FHR analysis. Each method is associated with details on application as well as hypothesis of physiological correlates to mechanisms performing control of the FHR signal.

Table 4.1 Analysis techniques in FHR

Method	Parameters	Length	Hypothesis
Frequency analysis AR model estimation from data and measurement of spectral components in defined frequency bands.	% of PSD (msec²) in frequency bands: LF: 0.03–0.15 Hz MF: 0.15–0.5 Hz HF: 0.5–1 Hz LF/(MF + HF)	3 min 360 values	Quantification of the activity of the autonomic nervous system.
Time analysis: morphological HR modification and variability	STV (msec) 7000II	70000 min 1200 values	Variability in the short period
	Delta FHR avg (msec) LTI (msec)	3 min 360 values	Variability in the long period
Approximate Entropy and sample entropy	ApEn (m,r,N) SampEn (m,r,N)	3 min 360 values	Presence of recurrent patterns in a single scale
Multiscale entropy	MSE: entropy estimator as a function of a scale factor τ_{sf}	7000 values	Investigation of signal structure: repetitive patterns are present at different time scales
Detrended fluctuation analysis (DFA)	DFA scaling exponent: linear relationship between the detrended integrated time series F(n) and the time scale on a log-log plot	1200 values	Quantification of long-range correlations in time series. Presence of power law (fractal) scaling Differences activity vs. quiet state
Lempel Ziv complexity (LZC)	LZC binary coding	7000 values	Rate of new patterns arising with the evolving of the signal
Phase-rectified signal averaging (PRSA)	APRS, DPRS	3–30 min	Detect quasi-periodicities in noisy, non-stationary signals

4.6 Application of Linear and NonLinear Analysis: Example of IUGR Detection

As an example of this novel solution in the monitoring of FHR, we present results in two selected populations: normal fetuses and i–tra-uterine growth restricted (IUGR) fetuses. IUGR is a metabolic dysfunction which does not allow the fetus to achieve its genetically predetermined size. The fetus is at risk of hypoxia and this condition is often associated with increased perinatal mortality and morbidity (morbidity 8%, incidence 4–7% in general population (15–20% in twins)). IUGR condition can bring to consequences either during pregnancy, till the fetal death and in the neonatal life with increased difficulties for the baby to achieve a normal development.

Monitoring the fetal growth, early in pregnancy, is essential, but the interpretation of clinical data in the very preterm period is very difficult and evidence-based guidelines are lacking. Even indices coming from new techniques (echo, Doppler fluximetry, etc.) did not solve the problem. As a matter of fact, the introduction of non-invasive ultrasound techniques allows assessing with high resolution the fetal biometry, i.e., identify the small fetuses. However a small fetus is not necessarily an IUGR fetus, but it can be only an SGA (small for gestational age). In this last case the fetus is small, but perfectly healthy unless its anatomic parameters are smaller than expected for its gestational age. SGA, in fact, is defined for the fetus with a biometric size <10° percentile. It would be very important to classify physiological cases of reduced dimensions from really pathological situations (IUGRs).

Cases of intrauterine growth restriction (IUGR) require different attention and management. There are no therapies and no definitive guidelines in IUGR. The prenatal management is aimed primarily at determining ideal timing and mode of delivery. Furthermore, no agreement was found in the definition of what is the expected growth potential for a fetus and who can be considered pathologically small.

This paragraph has the goal to provide an overview on how a set of quantitative indices from FHR analysis obtained from cardiotocographic recordings, during prenatal monitoring, can identify the fetal status during pregnancy including the IUGRs, separating them from the healthy SGAs.

The core summary of our contribution to the classification of the FHR signal can be described by some key steps. The first point is certainly the identification of a set of indices strictly related to the physiological mechanisms responsible for heart rate control in the fetus. The second step is to obtain relationships between these parameters and the condition of the fetus (normal or IUGR) through a multivariate approach. The physio–ogical-based features are then used as input of a simple multivariate classification model to early identify IUGR conditions during the antepartum period. All these findings open the possibility to introduce these novel estimators in the clinical practice.

The features extracted from methods previously described showed their capability in classifying different variability patterns in FHR.

The history of our analysis began with the application of frequency analysis methods in normal fetuses. This allowed us to verify that physiological development during gestation is marked by increasing fetal nervous system maturity which is measurable in the patterns of HRV signal. Even if the ANS is still developing its activity can be enhanced through spectral analysis parameters.

Low frequency, movement frequencies, and high frequency power components quantify the sympathetic and parasympathetic effect on the fetal heart regulation.

Here, we want to underline how features describing FHR variability can open new routes toward clinical evaluation. Tracking healthy fetus development is the first step to design an effective monitoring system [9].

Results were translated in an updated version of the 2CTG2 software.

The same approach was used for ApEn. Once it has been established that this entropy estimator was able to capture differences in fetal variability, these results opened to the use of this novel parameter in FHR analysis. The 2CTG2 software was upgraded with ApEn tool as this feature became well known even in the clinical obstetric world.

By the development of the research in nonlinear classifiers, novel parameters improving the entropy estimation were proposed. Sample entropy improved the consistence of the ApEn estimator, thus providing more robust measurements.

Comparison among different entropy estimators confirmed the usefulness of these features even in the classification of IUGR pathological fetal conditions [28]. The feature set produced by old and novel approaches provided multiparameter settings of FHR by different perspectives (time, frequency, complexity, regularity, etc.)

Differences among healthy and IUGR were enhanced by different parameters as the pregnancy progressed.

A step forward was moved by applying the entropy estimators (ApEn and SampEn) in a multiscale context.

Not only the variability of the FHR contains useful information that can be captured by regularity and complexity estimators, but these features can be calculated even at different scales (which means to consider coarse-grained versions of the original FHR signal).

The paper by [35] shows how MSE applied to a population of fetuses including normal, not severe, and severe IUGR provides a slope parameter at scale 1 and 2 that classifies the three groups as different each other.

Introduction of the LZC complexity [31] added a new and different index to FHR analysis. Even in this case, results show that a separation between healthy and IUGR fetuses is possible with statistic evidence.

New samples in FHR can bring new information and contribute to information enrichment in the signal. This enrichment can be connected to the development of an organized neural system that improved in time its control abilities till the birth and after it.

The role of physiological oscillation in FHR has always been considered the outmost importance for prenatal diagnosis. Acceleration and decelerations are signs of reactivity and sufferance, respectively, in most cases.

Phase-rectified signal averaging provides features that are related to the oscillatory component of the signal. Without hypothesis about periodicity PRSA analysis quantifies the incidence of positive (acceleration) and negative (deceleration) changes in the FHR signal.

APRS and DPRS parameters from phase-rectified signal averaging method contribute to the separation of healthy and IUGR fetuses with high statistical significance.

One reason for their performance is they are almost uncorrelated to other classical features in FHR analysis. This result reinforces the goodness of choosing features measuring different aspects of the FHR signal.

Once again, oscillations are mediated by nervous autonomic control as well as breathing-like influences. The optimal performance of these features is an indirect confirmation of their pathophysiological complex connections.

By looking at the performances of each single feature in the classification trial, results confirm an appreciable capability of each single parameter in separating fetal classes of healthy vs IUGR subjects [34].

However, in the recent years, it has become clear that a single index, although representative of a complex behavior, cannot be descriptive of all pathophysiological processes of pregnancy by itself, either it being computed in time, frequency, or nonlinear domain.

Despite the large availability of FHR quantitative indicators, very few fetal-related literature addressed the issue of how to adopt multivariate approaches in fetal surveillance. In the past, the availability of data was scarce. Recent years instead have seen a formidable growth in data generated during the care process. Moreover, data from fetal monitoring benefit from both technological advancements and novel available parameters which generate data increase.

As this chapter clearly illustrated, new features contribute to a better understanding of the fetal physiological system. They are also able to distinguish healthy from disease states. What is clear also is the increase in the computational complexity. Ranking and interpreting the results become a difficult job.

Thus, the need for a multimodal and multivariate analysis of FHR emerged as evident. Such novel integration should start from consolidated parameters toward the aim of integrating indices in a more comprehensive framework.

Our research group already proposed a bivariate approach to discriminate between healthy and IUGR fetuses [36]. Despite the interesting results, a two-dimensional space did prove not to be sufficient and it actually limits the investigation of a complex pathological condition on a reduced subset of features.

In this context, machine learning methods were adopted as their capability to find relationship in large and complex datasets. By submitting a large amount of raw data to ML classifiers, a low dimension model can be obtained explaining the largest amount of variance in the process and providing classification results.

This analysis was designed to search for an optimal decision rule in the multidimensional space of the parameters to predict the class of interest, namely healthy and IUGR. The classification rule is usually seen as a discrimination surface separating the multidimensional space into regions with homogeneous classes.

Different multivariate models were explored to identify several informative predictors of IUGR condition and to ultimately compare their discriminative performances: logistic regression (LR) [37], logistic regression stepwise (LR-SW) [38], naïve Bayes (NB) [39], classification trees (CT) [40], random forests (RF) [41], and support vector machines (SVM) [42].

The two machine learning techniques which outperformed, showing the best discriminative performances, were: RF (mean CA = 0.855, 95% CI = 0.794–0.916) and LR-SW (mean CA = 0.833, 95% CI = 0.759–0.908) which showed the best classification accuracy (CA) among the proposed machine learning models [43]. Although this study was carried out on a limited population (60 normals and 60 IUGRs), it constitutes a proof of concept that multivariate analysis on a set of physiopathological parameters can face complex diagnostic problems as the early identification of IUGRs and should be introduced on large scale in the clinical practice.

Once again, having some features in FHR which are linked to physiological mechanisms helps the interpretation and can guide the consequent clinical intervention.

This is the reason why, in our study, a two-step methodology for the early identification of the intrauterine growth restriction (IUGR) has been implemented, by deriving features from antepartum CTG traces. Methods were advanced signal analytics. The different machine learning techniques were trained with physiology-based FHR features: it made available a tool capable of providing an interpretable link between the machine learning results and the physiological mechanisms of fetal regulation.

4.7   Conclusions

As already pointed out at the beginning of the chapter, the aim was to provide the reader with information on the computerized systems for reading the CTG tracing, so that on the one hand there are no false expectations or illusions of absolute effectiveness, and on the other hand that techniques and numerical methods are correctly used.

The possibility of quantifying the analysis of the FHR through information technology and methodologies has allowed to overcome the main problem that plagued the traditional cardiotocographic technique, which is the subjective interpretation of the layout. The development of reliable algorithms for the identification of the various linear and time domain parameters extracted from the FHR signal has introduced the use of this approach in the clinical practice, even if the simple quantification of morphological analysis criteria (quantitative reproduction of the analysis by eye inspection) is not sufficient to highlight complex FHR alteration phenomena.

An advanced analysis that uses parameters more tied to a physiopathological meaning and more sophisticated classification methodologies, in addition to the advantage of providing objectivity and reproducibility to the cardiotocographic method, allows to highlight and study variation phenomena of the FHR correlated with the control function of the autonomic nervous system on local cardio-regulatory mechanisms. The possibility of being able to access these new assessments must be taken into consideration when assessing the importance of a computerized approach.

Furthermore, the use of classification tools, such as machine learning approaches, can lead to a further increase in the effectiveness of FHR analysis in the near future. The ability to learn general behaviors from available examples makes them particularly suitable for finding relationships that are not evident within sets of data, when causal knowledge is not well defined.

It is clear that it would be absolutely wrong from a clinical point of view to rely on the only abnormal result of a computerized CTG examination to decide, for example, the completion of childbirth in an IUGR fetus, but the global situation of the fetus and the mother must be considered; in fact, the data deriving from the computerized analysis of the CTG is a single clinical data and must be entered in a wider context.

In conclusion, the development of techniques and numerical methods for CTG analysis can constitute the future of fetal monitoring, especially for the early and reliable identification of states of suffering thanks to the integration of new knowledge of maternal-fetal pathophysiology with the extraction of significant parameters and with advanced classification methods.

References

1. Hammacher, K., Werners, P.H.: Über die auswertung und dokumentation von ctg-ergebnissen. Gynecol. Obstet. Investig. **166**(5), 410–423 (1968)
2. Alfirevic, Z., Devane, D., Gyte, G., Cuthbert, A.: Continuous cardiotocography (CTG) as a form of electronic fetal monitoring (EFM) for fetal assessment during labour. Cochrane Database Syst. Rev. **2**, CD006066 (2017). https://doi.org/10.1002/14651858.CD006066.pub3. ISSN 1469-493X. PMC 6464257. PMID 28157275.
3. Ayres-de-Campos, D., Bernardes, J.: FIGO Subcommittee. Twenty-five years after the FIGO guidelines for the use of fetal monitoring: time for a simplified approach? Int. J. Gynaecol. Obstet. **110**, 1):1–1):6 (2010)
4. Figueras, F., Albela, S., Bonino, S., et al.: Visual analysis of antepartum fetal heart rate tracings: inter- and intra-observer agreement and impact of knowledge of neonatal outcome. J. Perinat. Med. **33**, 241–245 (2005)
5. Grivell, R.M., Alfirevic, Z., Gyte, G., Devane, D.: Antenatal cardiotocography for fetal assessment. Cochrane Database Syst. Rev. (9), CD007863 (2015). https://doi.org/10.1002/14651858. CD007863.pub4. PMC 6510058. PMID 26363287
6. Lawson, C.W., Belcher, R., Dawes, G.S., Redman, C.W.G.: A comparison of an ultrasound (with autocorrelation) and direct electrocardiogram fetal heart rate detector system. Am. J. Obstet. Gynecol. **147**, 721–722 (1983)
7. Hewlett-Packard, G.H.: Fetal Monitor Test – a brief summary. Hewlett-Packard Technical Notes, Germany, pp. 1–6 (1995)
8. Chang, A., Sahota, D.S., Reed, N.N., James, D.K., Mohajer, M.P.: Computerised fetal heart rate analysis in labour-effect of sampling rate. Eur. J. Obstet. Gynecol. Reprod. Biol. **59**(2), 125–129 (1995)

9. Signorini, M.G., Magenes, G., Cerutti, S., Arduini, D.: Linear and nonlinear parameters for the analysis of Fetal heart rate signal from Cardiotocographic recordings. IEEE Trans Biom Eng. **50**(3), 365–374 (2003). https://doi.org/10.1109/TBME.2003.808824

10. Lowery, C.L., Russell Jr., W.A., Baggot, P.J., Wilson, J.D., Walls, R.C., Bentz, L.S., Murphy, P.: Time quantified detection of fetal movements using a new fetal movement algorithm. Am. J. Perinatol. **14**(1), 7–12 (1997)

11. Magenes, G., Signorini, M.G., Ferrario, M., Lunghi, F.: 2CTG2: A new system for the ante-partum analysis of fetal heart rate. In: 11th Mediterr. Conf. Med. Biomed. Eng. Comput. 2007, pp. 781–784. Springer Berlin Heidelberg, Berlin, Heidelberg (2007). https://doi.org/10.1007/978-3-540-73044-6_203

12. Macones, G.A., Hankins, G.D.V., SpongCY, H.J., Moore, T.: The 2008 National Institute of Child Health and Human Development workshop report on electronic Fetal monitoring. Obstet. Gynecol. **112**(3), 661–666 (2008). https://doi.org/10.1097/AOG.0b013e3181841395. PMID 18757666

13. Dawes, G.S., Houghton, C.R.S., Redman, C.W.G.: Baseline in human fetal heart rate records. Br J Obstet Gynecol. **89**, 270–275 (1982)

14. Mantel, R., Van Geijn, H.P., Caron, F.J.M., Swartjes, J.M., van Worden, E.E., Jongsma, H.W.: Computer analysis of antepartum fetal heart rate: I. Baseline determination. Int. J. Biomed. Comput. **25**, 262–272 (1990)

15. Bailey, R.E.: Intrapartum fetal monitoring. Am. Fam. Phys. **80**(12), 1388–1396 (2009). PMID 20000301.

16. Arduini, D., Rizzo, G., Piana, G., Bonalumi, A., Brambilla, P., Romanini, C.: Computerized analysis of fetal heart rate: I. Description of the system (2CTG). J. Matern. Fetal Invest. **3**, 159–163 (1993)

17. Task Force of the Europ. Soc. of Cardiol. & North Am. Soc.of Pacing and Electrophys: Heart rate variability, standard of measurement, physiological interpretation and clinical use. Circulation. **93**, 1043–106521 (1996)

18. Cerutti, S., Civardi, S., Bianchi, A., Signorini, M.G., Ferrazzi, E., Pardi, G.: Spectral analysis of antepartum heart rate variability. Clin. Phys. Physiol. Meas. **10 Suppl B**, 27–31 (1989)

19. Kristal-Boneh, E., Rafeil, M., Froom, P., Ribak, J.: Heart rate variability in health and disease. Scand. J. Work Environ. Health. **21**, 85–89 (1995)

20. Divon, M.Y., Zimmer, E.H., Yeh, S.Y., Vilensky, A., Sarna, E.Z., Paldi, E., Platt, L.D.: Modulation of foetal heart rate patterns by foetal breathing: the effect of maternal intravenous glucose administration. Am. J. Perinatol. **2**, 292–295 (1985)

21. Giuliano, N., Annunziata, M.L., Esposito, F.G., Tagliaferri, S., Di Lieto, A., Magenes, G., Signorini, M.G., Campanile, M., Arduini, D.: Computerised analysis of antepartum foetal heart parameters: new reference ranges. J. Obstet. Gynaecol. **37**(3), 296–304 (2017)

22. Magenes, G., Signorini, M.G., Arduini, D., Cerutti, S.: Fetal heart rate variability due to Vibroacoustic stimulation: linear and nonlinear contribution. Methods Inf. Med. **43**, 47–51 (2004)

23. Lecanuet, J.P., Schaal, B.: Fetal sensory competencies. Eur. J. Obstet. Gynecol. **68**, 1–23 (2006)

24. Marple, L.: Digital Spectral Analysis with Applications, pp. 264–285. Prentice-Hall, Englewood Cliffs (1971)

25. Pincus, S.M.: Approximate entropy (ApEn) as a complexity measure. Chaos. **5**, 110–117 (1995)

26. Richman, J.S., Moorman, J.R.: Physiological time-series analysis using approximate entropy and sample entropy. Am. J. Physiol. Heart Circ. Physiol. **278**, 2039–2049 (2000)

27. Costa, M., Goldberger, A.L., Peng, C.K.: Multiscale entropy analysis of complex physiologic time series. Phys. Rev. Lett. **89**, 068102 (2002)

28. Ferrario, M., Signorini, M.G., Magenes, G., Cerutti, S.: Comparison of entropy-based regularity estimators: application to the fetal heart rate signal for the identification of fetal distress. I.E.E.E. Trans. Biomed. Eng. **53**(1), 119–125 (2006)

29. Signorini, M.G., Magenes, G., Ferrario, M., Lucchini, M.: Complex and nonlinear analysis of heart rate variability in the assessment of fetal and neonatal wellbeing. In: Complexity

and Nonlinearity in Cardiovascular Signals, Springer international Publishing AG, Cham, Switzerland pp. 427–450 (2017)

30. Lempel, A., Ziv, J.: On the complexity of finite sequences. IEEE Trans. Inf. Theory. **22**, 75–81 (1976)

31. Ferrario, M., Signorini, M.G., Magenes, G.: Comparison between fetal heart rate standard parameters and complexity indexes for the identification of severe intrauterine growth restriction. Methods Inf. Med. **46**, 186–190 (2007). doi:07020186

32. Bauer, A., Bunde, A., Barthel, P., Schneider, R., Malik, M., Schmidt, G.: Phase rectified signal averaging detects quasi-periodicities in nonstationary data. J. Phys. A. **364**, 423–434 (2006)

33. Fanelli, A., Magenes, G., Campanile, M., Signorini, M.G.: Quantitative assessment of fetal well-being through ctg recordings: a new parameter based on phase-rectified signal average. IEEE J. Biomed. Heal. Inform. **17**(5), 959–966 (2013)

34. Signorini, M.G., Fanelli, A., Magenes, G.: Monitoring fetal heart rate during pregnancy: contributions from advanced signal processing and wearable technology. Comput. Math. Methods Med. **2014**, 707581 (2014)

35. Ferrario, M., Signorini, M.G., Magenes, G.: Complexity analysis of the fetal heart rate variability: early identification of severe intrauterine growth-restricted fetuses. Med. Biol. Eng. Comput. **47**(9), 911–919 (2009)

36. Ferrario, M., Signorini, M.G., Magenes, G.: Complexity analysis of the fetal heart rate for the identification of pathology in fetuses. In: Computers in Cardiology, 32, art. no. 1588275, IEEE Publisher, Piscataway, USA. pp. 989–992 (2005)

37. McCullagh, P., Nelder, J.A.: Generalized Linear Models. Chapman and Hall, Chapman & Hall/CRC, Boca Raton, Florida, USA(1989)

38. Bursac, Z., Gauss, C.H., Williams, D.K., Hosmer, D.W.: Purposeful selection of variables in logistic regression. Source Code Biol. Med. **3**, 17 (2008). https://doi.org/10.1186/1751-0473-3-17

39. Hand, D.J., Yu, K.: Idiot's Bayes – not so stupid after all? Int. Stat. Rev. **69**, 385–398 (2001). https://doi.org/10.1111/j.1751-5823.2001.tb00465.x

40. Kotsiantis, S.B.: Decision trees: a recent overview. Artif. Intell. Rev. **39**, 261–283 (2013). https://doi.org/10.1007/s10462-011-9272-4

41. Breiman, L., Friedman, J.H., Olshen, R.A., Stone, C.J.: Classification and Regression Trees. Chapman & Hall/CRC, Boca Raton, Florida, USA (1984)

42. Zhang, Y.: Support Vector Machine Classification Algorithm and its Application, vol. 2012, pp. 179–186. Springer, Berlin, Heidelberg (2012). https://doi.org/10.1007/978-3-642-34041-3_27.35

43. Signorini, M.G., Pini, N., Malovini, A., Bellazzi, R., Magenes, G.: Integrating machine learning techniques and physiology based heart rate features for antepartum fetal monitoring. Comput. Methods Prog. Biomed. **185**, 105015 (2020)

Chapter 5
Noninvasive Fetal Electrocardiography: Models, Technologies, and Algorithms

Reza Sameni

Contents

R. Sameni (✉)
Department of Biomedical Informatics, Emory University School of Medicine,
Atlanta, GA, USA
e-mail: rsameni@dbmi.emory.edu; http://www.sameni.info/

© Springer Nature Switzerland AG 2021
D. Pani et al. (eds.), *Innovative Technologies and Signal Processing in Perinatal Medicine*, https://doi.org/10.1007/978-3-030-54403-4_5

5.1 Introduction

The early assessment of fetal well-being is the major objective of fetal monitoring during pregnancy and labor. The latter is specifically useful for identifying fetuses at risk of hypoxia (oxygen deficiency) during labor. In this context, fetal electrocardiography is one of the emerging technologies, which dates back to 1906 [27] but has gained much more attention during the past two decades. The technology has significantly evolved throughout the past 50 years, from naive visual inspection to multichannel automatic methods of noninvasive fetal electrocardiogram (fECG) extraction, using advanced signal processing methods [41, 77]. The method has become more popular in recent years due to its relatively low cost and advances in the required signal acquisition and signal processing techniques. In this context, both invasive methods used after amniotic sac rupture during labor and noninvasive methods using maternal abdominal leads throughout pregnancy (especially during the third trimester) have been used. Although invasive fECG recording using fetal scalp leads has a higher signal-to-noise ratio (SNR) and requires less processing as compared with noninvasive signals captured from the maternal abdomen, due to the potential risks of invasive methods for both the mother and the fetus(es), it is not so popular. On the other hand, despite its advantages, noninvasive fECG extraction is hampered by many practical challenges including (1) the significantly lower SNR of the fECG as compared with the maternal ECG (mECG), which superposes over the abdominal leads; (2) device and measurement issues related to noninvasive fECG acquisition using single or multiple maternal abdominal sensors; (3) the indirect access to the fetal heart through multiple maternal body layers, which act as a *volume conductor*; (4) artifacts and variations in fECG shape due to fetal movements; (5) baseline wanders of the data due to maternal respiration; and (6) measurement and environmental noises such as maternal muscle and uterine contractions, power-line noise, and artifacts due to other bedside monitors and devices such as the infusion pumps. Most of these noises overlap with the fECG in time, frequency, and space (leads), making fECG extraction a nontrivial challenge, which requires advanced signal processing.

To date, various methods have been developed for fECG extraction with various degrees of success, including adaptive filtering [7, 32, 56, 59, 67, 69, 89, 97, 104], Kalman filtering [65, 82, 84], singular value decomposition [45], blind and semi-blind source separation using independent and periodic component analyses [28, 74, 83, 108], and wavelet transforms [47, 55, 101]. Some of these techniques, such as Kalman filters, singular value decomposition, wavelets, and adaptive filters (used in line-enhancement mode) have been applied to both single and multichannel abdominal ECG recordings. In contrast, other techniques such as independent component analysis or adaptive noise cancellation using an external reference require two or more channels of measurements. Multichannel techniques based on blind and semi-blind source separation have proved to be very effective to overcome the aforementioned challenges. Nevertheless, various aspects of noninvasive fECG extraction are still open problems and require further studies—for example, issues related to long-time online fECG monitoring (required for fetal Holter monitoring), problems due to fetal movements during signal acquisition, variations in fECG morphology (again due to fetal motion and fetal positioning with respect to the body surface leads), and fECG extraction in low SNR using few numbers of channels. There are also several post-fECG extraction issues including fetal R-peak detection, heart rate (HR) calculation, fECG morphology extraction, and clinical parameter extraction (QT interval, ST-level calculation, etc.) from noisy fECG signals. From the clinical and industrial perspective, the size and cost of the device, the technology, and the number of maternal abdominal leads (preferably only a few leads placed close together in a patch of electrodes) are also of great importance.

In this chapter, the major signal processing techniques, which have been developed for the modeling, extraction, and analysis of the fECG from noninvasive maternal abdominal recordings over the past 50 years, are reviewed and compared with one another in detail.

5.2   Noninvasive Fetal Electrocardiography Data Model

5.2.1   Volume Conductor Model

The physics of the problem of noninvasive fECG measurement from the maternal abdomen follows the general principles of volume conduction theory [43]. The properties of the propagation media from the fetal heart to the maternal abdomen have been explored in previous studies [66, 74]. The major aspects of the problem, which influence the fECG data model and extraction techniques, can be summarized as follows [41]:

1. *Negligible electric displacement current*: The electromagnetics of the problem is quasi-static. Therefore, the electric and magnetic fields are decoupled, the electric field is proportional to the gradient of the electric scalar potential, and the divergence of the current density is zero.
2. *Linear propagation media*: Superposition holds for the electrical potentials due to the maternal heart, fetal heart, and other sources of biopotentials.
3. *Negligible capacitive component of the body tissues' electrical impedance*: Due to the relatively low frequency range of interest (below 10 kHz), the tissues are to a very good approximation resistive and the capacitance is negligible.
4. *Spatial distribution of the heart*: The source signals are *non-punctual*, and different lead configurations provide different views of the heart, conveying different— although rather redundant and correlated— information. Therefore, the cardiac source may only be approximated by a current dipole in the far-field.
5. *Non-homogeneous volume conductor*: Low-conductivity layers, such as the *vernix caseosa*, which form throughout pregnancy (mainly between weeks 28 and 32 of gestation [77]), can change the preferred electrical propagation pathways, resulting in morphological variations on the maternal body surface [66, 94].

6. *Morphological variability*: During a signal recording session, although the fECG morphology is consistent with respect to the fetal body (as in adult ECG)—due to fetal motions such as rotations, movements of extremities, and hiccups—the extracted fECG morphology can change with respect to the maternal body coordinate system and the maternal body surface sensors. Moreover, minor fetal and maternal movements, such as maternal respiration, somehow *modulate* the fetal cardiac signals acquired from the maternal abdomen.

These properties imply that temporal parameters such as the R-peak locations, heart rate, and PT and QT intervals can be very accurate, but parameters, such as the R-wave amplitudes and T-to-R ratios, which rely on amplitudes and ratios of amplitudes are totally unreliable since they can easily change with fetal positioning, gestation age, or a change of lead configurations. Nevertheless, relative variations of amplitude-based parameters can still be accurate between successive fetal heart beats and during real-time monitoring. For example, phenomena such as *T-wave alternans* (TWA) which require the comparison of the T-wave amplitudes between successive beats are still reliable (up to the signal quality).

Note that items 1–4 listed above are also applicable to adult ECG and the mECG that superposes over the abdominal leads. Based on these properties, the problem of noninvasive fECG acquisition from an array of maternal abdominal sensors can be mathematically formulated as follows:

$$\mathbf{x}(t) = \mathbf{H}_m \mathbf{s}_m(t) + \mathbf{H}_f \mathbf{s}_f(t) + \mathbf{H}_v \mathbf{v}(t) + \mathbf{n}(t) \qquad (5.1)$$

where $\mathbf{x}(t) \in \mathbf{R}^n$ is the n channel of maternal body surface measurements acquired differentially with respect to one or more reference channels, $\mathbf{s}_m(t) \in \mathbf{R}^m$ is the mECG source component, $\mathbf{s}_f(t) \in \mathbf{R}^l$ is the fECG source component, $\mathbf{v}(t) \in \mathbf{R}^k$ represents the structured (correlated or low-rank) noise corresponding to other biopotential sources (such as maternal muscle contractions) or device noise, and $\mathbf{n}(t) \in \mathbf{R}^n$ is the unstructured (full-rank) measurement noise, which corresponds to sensorwise noise that is uncorrelated from the other signals and structured noises. In the data model (5.1), $\mathbf{H}_m \in \mathbf{R}^{n \times m}$, $\mathbf{H}_f \in \mathbf{R}^{n \times l}$, and $\mathbf{H}_v \in \mathbf{R}^{n \times k}$ are the *lead-field* matrices, which map the source components to the body surface electrode recordings. The model may be further extended to consider minor maternal body motions (e.g., due to respiration) and fetal movements by considering $\mathbf{H}_m$, $\mathbf{H}_v$, and $\mathbf{H}_f$ to be the functions of time. Also in multiple pregnancies, similar terms can be added to (5.1) for the other fetuses [81].

The spatial distribution of the cardiac source implies that in (5.1), m and l theoretically tend to be infinity. However, as we get farther from the cardiac sources, far-field approximations are applicable and the cardiac sources behave more like dipoles [58]. Therefore, in practice, each of the cardiac sources can be approximated up to finite *effective number of dimensions* [88]. In [80], it was quantitatively shown that for adult ECG, taking m between 5 and 6, and for fetal ECG, assuming l between 1 and 3, are sufficient to retrieve the major energy fraction of the maternal and fetal ECG components (from the maternal abdominal lead recordings).

Apparently, the effective number of dimensions also depends on the sensor position with respect to the maternal and fetal hearts. For example, if the maternal abdominal leads are placed rather distanced from the maternal chest, or if the fetal position is such that the shortest conductive path between the differential sensor pairs does not pass through the fetal heart (i.e., the fetal cardiac electrical fields do not result in significant potential differences between the recording differential pair leads), the effective number of dimensions reduces. In this case, the fECG is not retrievable from the abdominal leads, even by using the most advanced signal processing techniques. It is later shown that the effective number of dimensions and the number of maternal body sensors are specifically important for multichannel fECG extraction algorithms. Some general guidelines for selecting the sensor locations for better fECG retrieval are presented in Sect. 5.3.3.

5.2.2 Morphological Model

5.2.2.1 Template-Based Models

Mathematical modeling of the ECG waveform has vast applications in ECG device test instruments and for educational purposes. To date, the beat-wise ECG morphology has been modeled by various mathematical functions including Bessel functions [93], Hermite polynomials [48], and Gaussian functions [39, 61]. The latter has an intrinsic dynamic mechanism for generating continuous ECG waveforms, which will be later discussed in details. Other wave-based models can generate a continuous ECG by replicating a fixed waveform that resembles the beat-wise ECG morphology. Accordingly, a single-channel ECG can be modeled as follows:

$$\mathrm{ecg}(t) = \sum_n h(t - T_n; \gamma_n), \quad T_n = nT + \eta_n \tag{5.2}$$

where T_n denotes the R-peak locations, T is the average RR-interval, η_n is the RR-interval deviation, $h(t; \cdot)$ is the ECG morphology, and γ_n denotes the beat-wise variations of the ECG morphology considered as a model parameter. It is shown in the sequel that this simple pseudo-periodic model can be used for removing mECG interferences from the fECG. The limitation of this model is that the natural beat-wise variations of the heart rate, which result in the shortening or prolongation of certain segments of the ECG, are not explicitly considered in this model. In fact, a more accurate model should permit the compression and expansion of the ECG morphology, as the heart rate evolves over time. Based on this requirement, the notion of *cardiac phase* has been introduced for modeling and development of ECG filtering and later used for mECG cancellation and fECG extraction from multi-channel abdominal recordings.

5.2.2.2   The Notion of Cardiac Phase

The cardiac cycle, or the period from one sinoatrial (SA) node activation to the next, consists of a period of relaxation (diastole), during which the heart is filled with blood, followed by a period of contraction (systole), as shown in Fig. 5.1. For a normal heart, the contraction and relaxation phases are subject to continuous change, controlled by the autonomic nervous system, and these changes do not necessarily take place "linearly" along the beats. In other words, when the heart rate changes, the different segments of the ECG are not scaled to the same extent. Specifically, it is believed that when the heart rate increases, e.g., due to physical activity, tachycardia, and bradycardia, the duration of the action potentials and the period of the systolic phase also decrease, but not as much as the variations of the diastolic phase of the ECG [36]. Alternation in the cardiac cycle duration depends on various physiological factors, which can be modeled using the notion of *cardiac phase*. As proposed in [83], the cardiac phase $\theta(t) \in [-\pi, \pi]$ (or alternatively $[0, 2\pi]$) can be used as a variable for the mathematical representation of the pseudo-periodic behavior of the heart over different beats. As illustrated in Figs. 5.2 and 5.3, each electrophysiological state of the heart over a full cardiac cycle can be mapped to a unique value between $[-\pi, \pi]$. In other words, the linear phase $\theta(t)$ provides a means of phase-wrapping the RR-interval onto the $[-\pi, \pi]$ interval. Therefore, the ECG— regardless of its RR-interval deviations— is converted to a polar representation, in which the ECG components in different beats, such as the P, Q, R, S, and T-waves, are more or less phase-aligned with each other, especially over the QRS segment (Fig. 5.4). As a result, identical contraction or relaxation states of the heart are mapped to identical values of $\theta(t)$. For example, by convention, the peak of the systole (the R-peak) can be fixed to $\theta(t) = 0$. This convention maps the *ventricular diastolic* state of the heart to negative phases and the *ventricular systolic* state to positive phases. In this case, the phase-wrapping from $-\pi$ to π takes place just after

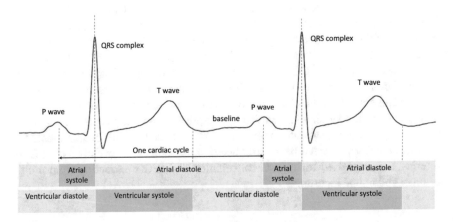

Fig. 5.1 The cardiac states across successive beats versus the ECG

Fig. 5.2 The cardiac cycle phase-wrapped on the unit circle using the phase signal. The heart sounds S1 and S2 are also demonstrated for reference to the mechanical activity of the heart

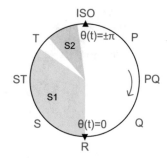

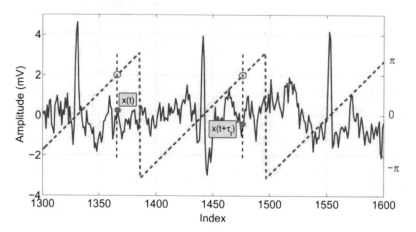

Fig. 5.3 The cardiac phase using a linear phase. (Adapted from [83])

the T-wave offset, and at the beginning of the relaxation period of the heart, where the ECG level is at its isoelectric point or baseline (cf. Figs. 5.1 and 5.2).

From the cardiac phase signal, some other quantities can be calculated, which have been extensively used in the literature, for modeling and denoising adult and fetal ECG signals:

- *Cardiac angular frequency and instantaneous heart rate:* The *cardiac angular velocity* $\omega(t)$ in rad/s and the *instantaneous heart rate* in Hz are defined as follows:

$$\omega(t) = 2\pi f(t) = \frac{d\theta(t)}{dt} \tag{5.3}$$

- Therefore, the conventional RR-interval can be considered as the average of the reciprocal of $f(t)$, over one beat. Note that both $f(t)$ and $\omega(t)$ are rather abstract quantities for conventional ECG analysis, in the sense that only the RR-interval is known as a clinical index (the duration between the onsets of successive

Fig. 5.4 Polar
representation of a noisy
ECG using the cardiac
phase signal $\theta(t)$ [74]

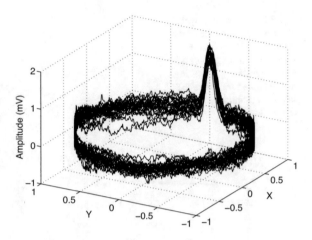

ventricular systoles). Nevertheless— again in an abstract sense—$f(t)$ and $\omega(t)$
can be considered as the speed of cardiac dipole rotation in the myocardium.

- *Time-varying cardiac period:* In each ECG cycle, the sample at the time instant
  t has a *dual sample* in other beats, which have the same phase value. We define
  the distance between sample t and its dual sample in the previous beat as the
  time-varying period, which is denoted by τ_t and mathematically defined as
  follows:

$$\tau_t = \arg\min_{\tau>0}\{\theta(t-\tau) = \theta(t)\} \qquad (5.4)$$

5.2.2.3 Dipolar Models

According to dipolar models of the heart [57, 58], the signals acquired from different body surface leads are projections of the cardiac dipole vector onto the recording electrode axes. Due to the properties of the fetal and maternal body volume
conductors, detailed in Sect. 5.2.1, the signals acquired by all body surface leads are
quasi-periodically time synchronous with the cardiac phase. These properties have
been used in the literature to develop synthetic models for generating maternal and
fetal cardiac waveforms. The first modeling framework, explicitly focused on the
fECG, was developed in [66]. This study was based on maternal body surface potentials modeling using finite elements methods and assuming a dipolar model for the
fetal heart. Another popular model is based on the single-channel ECG model proposed by McSharry and Clifford [18, 19, 61, 79], which was later extended to the
fECG in [81]. Accordingly, the following dynamic model has been proposed for
simulating the three dipole coordinates of the *vectorcardiogram* (VCG), which is
denoted by $\mathbf{s}(t) = [x(t), y(t), z(t)]^T$:

$$\dot{\theta} = \omega$$

$$\dot{x} = -\sum_i \frac{\alpha_i^x \omega \Delta\theta_i^x}{\left(b_i^x\right)^2} \exp\left[-\frac{\left(\Delta\theta_i^x\right)^2}{2\left(b_i^x\right)^2}\right]$$

$$\dot{y} = -\sum_i \frac{\alpha_i^y \omega \Delta\theta_i^y}{\left(b_i^y\right)^2} \exp\left[-\frac{\left(\Delta\theta_i^y\right)^2}{2\left(b_i^y\right)^2}\right] \tag{5.5}$$

$$\dot{z} = -\sum_i \frac{\alpha_i^z \omega \Delta\theta_i^z}{\left(b_i^z\right)^2} \exp\left[-\frac{\left(\Delta\theta_i^z\right)^2}{2\left(b_i^z\right)^2}\right]$$

where $\Delta\theta_i^x = \left(\theta - \theta_i^x\right)\mathrm{mod}\left(2\pi\right)$, $\Delta\theta_i^y = \left(\theta - \theta_i^y\right)\mathrm{mod}\left(2\pi\right)$,

$\Delta\theta_i^z = \left(\theta - \theta_i^z\right)\mathrm{mod}\left(2\pi\right)$, $\omega = 2\pi f$ are the cardiac angular velocities and f is the instantaneous heart rate, as defined in (5.3). Mathematically, the first equation in (5.5) generates a circular trajectory, which rotates with the frequency of the heart rate. In other words, each cycle of θ sweeping from 0 to 2π corresponds to one cardiac cycle, and the other equations model the dynamics of the three coordinates of the source vector $s(t)$ as a summation of Gaussian functions with amplitudes α_i^x, α_i^y, and α_i^z, widths b_i^x, b_i^y, and b_i^z, each located at rotational angles θ_i^x, θ_i^y, and θ_i^z. The intuition behind this set of equations is that the baseline of each of the dipole coordinates is pushed up and down, as the trajectory approaches the centers of the Gaussians, resulting in a moving vector in the (x, y, z) coordinate space. In practice, by adding some deviations to the parameters of (5.5), for example by considering them as random variables rather than deterministic constants, more realistic ECG with inter-beat variations can be generated.

The above model of the rotating dipole vector is rather general, since due to the *universal approximation* property of Gaussian mixtures, any continuous function such as the dipole vector coordinates can be modeled with a sufficient number of Gaussian functions, up to an arbitrarily close approximation [11]. Moreover, the model is a very good choice for ECG signals of both adults and fetuses, for which the Gaussian kernels can be eventually related to clinical parameters of the ECG. Equation (5.5) can also be thought as a model for the orthogonal lead VCG coordinates, with an appropriate scaling factor for the attenuations of the volume conductor. This analogy between the orthogonal VCG and the dipole vector was used in [81] to estimate the parameters of (5.5) from the three *Frank-lead* VCG recordings.

By placing the resulting cardiac source models of the maternal and fetal cardiac dipoles in (5.1), realistic mixtures of maternal abdominal signals are obtained. In Figs. 5.5 and 5.6, a sample signal corresponding to the cardiac dipole coordinates and the resulting three-dimensional vectorcardiogram loop are shown for illustration. A multichannel signal generated by this technique plus synthetic noise is also shown in Fig. 5.7. The functions required for generating synthetic maternal

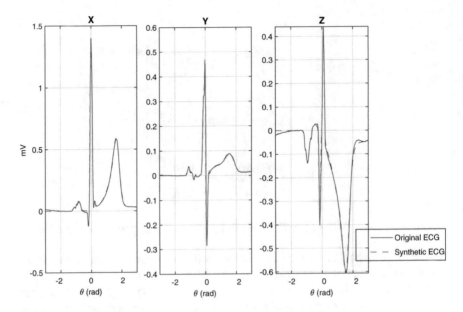

Fig. 5.5 Synthetic ECG signals generated by the VCG model in (5.5)

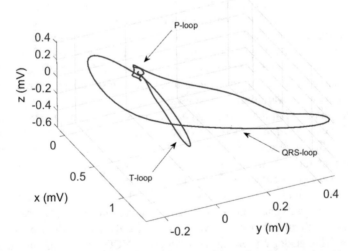

Fig. 5.6 Typical synthetic VCG loop. Each clinical lead is produced by mapping this trajectory onto a one-dimensional vector in this three-dimensional space

abdominal signals are available online in the website mentioned in [76], with the parameter set listed in [81]. Accordingly, the number of the Gaussian functions used for modeling the maternal and fetal ECG are not necessarily the same for the different channels and they can be selected according to the shape of the desired channel.

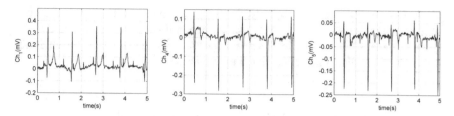

Fig. 5.7 Typical multichannel ECG generated by a synthetic maternal-fetal ECG generator

Databases of synthetic maternal and fetal cardiac signals generated by this method are available online for algorithm evaluation [2, 76].

5.3   Digital Noninvasive Fetal ECG Acquisition

5.3.1   Acquisition Front-End Requirements

To date, there are no standards or widely accepted protocols for fECG acquisition. Nevertheless, the common properties of the fetal and adult ECG and the existing open-access fECG databases can be used to set some baselines. It is known that the effective bandwidth of adult ECG is between 0.05 and 150 Hz, with a maximum span of ±5 mV in magnitude, besides the common-mode and electrode offset voltages, as shown in Fig. 5.8. It is recommended that the front-end noise of adult ECG devices should be below 30 μV in root mean square (RMS) [24]. On the other hand, in the currently available maternal abdominal datasets, the fECG can be 10–20 times smaller than the mECG. At the same time, due to the sharper QRS and higher heart rate of the fetus as compared with the adult ECG, the fECG is wider in bandwidth. As a baseline, a bandwidth between 0.05 and 250 Hz covers the dominant bandwidth of the fECG. In this range, the most informative band is from 10 to 70 Hz, which is used for fetal heart rate detection, while the full bandwidth is recommended for fECG morphological analysis.

According to the sampling theorem, the sampling frequency of a signal should be above twice the maximum frequency of the input signal (known as the *Nyquist rate*) to avoid *aliasing* and to guarantee information retrieval. But for biomedical applications, signal visualization is an integral aspect of the analysis, and sampling at the minimal Nyquist rate does not result in visually agreeable signals. Therefore, biomedical signals are commonly over-sampled above the Nyquist rate for better visualization and possible SNR improvement during post-processing.

As for the amplitude, fECG acquisition systems should have a broad dynamic range to permit fECG acquisition without overflow or saturation due to interfering signals such as the mECG and power-line noise, as demonstrated in Fig. 5.9. In Fig. 5.10, the amplitude and frequency range of the fECG are compared with other biosignals and artifacts. Accordingly, the fECG spectrally overlaps with the interfering biosignals and is significantly weaker in amplitude. Therefore, classical frequency domain filtering is ineffective, especially for the mECG, which is the dominant biomedical interfering signal for the fECG.

5.3.2   Analog-to-Digital Conversion Requirements

The procedure of analog-to-digital signal conversion inevitably adds quantization noise to the signal and reduces the signal-to-noise ratio (SNR). It is therefore important to keep the quantization noise below or at the same level as the analog signal noise level to avoid significant signal quality degradation. The SNR due to the quantization procedure can be calculated from the standard equation:

$$\text{SNR}(\text{dB}) = 6.02b + 1.76 + 10\log_2(\text{OSR}) \tag{5.6}$$

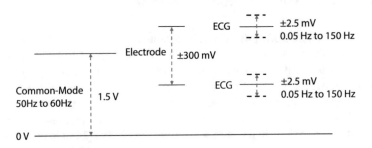

Fig. 5.8 The dynamic range of analog ECG frontends. (Adapted from [98])

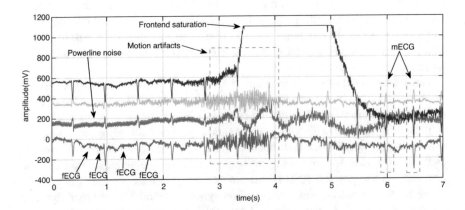

Fig. 5.9 A typical segment of maternal abdominal recordings containing various signals and noises. The dynamic range of the digital front-end should be such that the acquired signals would not overflow due to interfering signals such as the maternal ECG. Refer to the text for further details

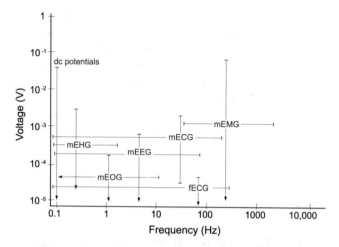

Fig. 5.10 The amplitude and frequency range of the maternal electrocardiogram (mECG), electro-encephalogram (mEEG), electrooculogram (mEOG), electromyogram (mEMG), electrohystro-gram (mEHG), and the fetal ECG (fECG). Accordingly, different biosignals interfere with the fetal ECG [31, 81, 91, 102]. Note that the fECG amplitude depends on the sensor position, fetal positioning, and age

where b is the number of analog-to-digital converter (ADC) bits and OSR $= f_s/$BW is the *over-sampling ratio*, which is the ratio of the sampling frequency f_s and the bandwidth (BW) of the input signal. The SNR improvement due to the OSR term in (5.6) is only obtained by post-filtering if the signal is sampled above the minimal Nyquist rate. Note that the standard SNR equation (5.6) is based on the assumption of a sinusoidal input signal with close- to full-scale amplitude range (typically 1 dB below the ADC full-scale level) applied to a symmetric voltage referenced ADC with uniform quantization levels and assuming that the quantization noise is uniformly distributed over the entire Nyquist bandwidth [46]. This standard procedure enables the manufacturers and circuit designers to have a unified comparison between different ADC devices.

It should also be noted that in digital electronics circuits design, the maximum SNR expected from the nominal number of ADC bits is not achievable. In fact, depending on the ADC technology, sampling frequency, and the printed circuit board (PCB) design and quality, the *effective number of bits* (ENOB) is what is obtained in practice:

$$\bar{b} = \frac{\text{SNR}_{\text{real}} - 1.76\text{dB}}{6.02} \tag{5.7}$$

where SNR_{real} is the SNR that is obtained in practice and $\bar{b}$ is the ENOB, which is not necessarily an integer value. For example, an ADC with 16 nominal bits may practically have 13.5–14 ENOBs. The ENOB is one of the standard properties of all ADC, which is documented in the datasheets of ADC devices by the manufacturers.

Considering that beyond the ADC chip technology the ENOB also depends on the circuit design quality, it is measured in practice by sweeping close- to full-scale sinusoidal signals within the Nyquist band of the manufactured circuit front-end (by applying a signal generator to the ADC front-end) and by logging the samples acquired by the ADC. The real SNR (SNR_{real}) can be eventually calculated by analyzing the sampled signals in software. This is a standard procedure that is performed during the design and quality control of all (including medical) equipment. The overall recommended front-end specifications for noninvasive fECG acquisition are summarized in Table 5.1.

5.3.3 Sensor Placement

In order to maximize the chance of retrieving the fECG from maternal abdominal leads, it is common to use multiple leads spread over the abdomen, lower back, and the two sides of the maternal body. The sensors should ideally be close to the fetus and the referencing of the leads should be such that the electrical fields due to the fetal heart pass through the differential pairs used for acquisition. To date, the number of abdominal channels used for research and clinical usage are very diverse, ranging from as few as one to as many as 144 abdominal channels. From the electronic and manufacturing perspective, using a few leads placed close together in a patch of disposable or reusable electrodes is very advantageous, as compared with using numerous electrodes distributed all over the maternal abdomen and back. However, as explained throughout this chapter, a group of sensors placed close to each other are prone to becoming highly dependent and result in mathematically low-rank and non-invertible mixture of signals, which is inappropriate for multi-channel fECG extraction. Therefore, there is a compromise between the simplicity

Table 5.1 The recommended front-end specifications for fetal ECG acquisition

Property	Range
Bandwidth (−3 dB cutoff frequency)	Acceptable: 0.05–250 Hz Preferred: 0.05 Hz to 1 kHz (for better fECG-noise separability)
Amplified analog voltage range	3–5 V (preferably differential pairs)
Analog-to-digital resolution	Low resolution: 16 bits High resolution: 24 bits
Sampling frequency	Minimum: 500 Hz Acceptable: 1 kHz High resolution: 5–10 kHz
Sampling sequence	Preferred: simultaneous Acceptable: sequential (multiplexed); only at high sampling frequencies
Number of channels	Between 8 and 32 with dedicated mECG channels used as reference

of the acquisition system and the robustness to fetal positioning. The major fECG acquisition technologies use between 8 and 32 channels, including one or more reference leads for the mECG acquired from maternal chest leads.

5.4 Single-Channel Fetal Electrocardiogram Extraction

Single-channel fECG extraction algorithms refer to the category of methods that use a single maternal abdominal channel and possibly a set of reference electrodes for acquiring the mECG from the maternal chest. An interesting comparative survey on the advantages and limitations of these methods was conducted in [8]. In this section, some of the major algorithms of this class of techniques are reviewed in further detail.

5.4.1 Naive Fetal Electrocardiogram Detection and Extraction

Before the advances in digital signal processing in recent decades, fECG detection was performed over raw paper prints of abdominal recordings, without any processing. For instance in [50], by visual inspection, several cases were reported in which due to the *vertex* presentation of the fetus, the fetal R-peaks appeared as positive peaks while the maternal R-peaks had negative peaks. It is evident that such studies remained discrete and subjective, since due to the low SNR, fECG detection by visual inspection is not always applicable and highly dependent on the fetal presentation and gestational age. Nevertheless, visual inspection remains as the first intuitive test for machine-based fECG extraction algorithms.

5.4.2 Template Subtraction and Cyclostationary Random Process Theory

Template subtraction is the most basic method for mECG cancellation from maternal abdominal recordings [1, 54]. Despite its simplicity, it was shown in [41] that using the theory of cyclostationarity, this technique can be the *optimal cyclostationary Wiener filter*, when applied properly by compensating the inter-beat variations of the mECG. The proof was inspired by the problem of pulse amplitude demodulation, a well-known method in the context of telecommunications [35, Ch. 4].

Let us consider the signal $x(t) = \sum_n c_n g(t - nT)$, where $g(\cdot)$ is an arbitrary *known* function and c_n is a stationary time-sequence. It can be shown that the problem of optimal filtering of $x(t)$, which is a *wide-sense cyclostationary* random process, from the additive mixture $z(t) = x(t) + \eta(t)$ (where $\eta(t)$ is a stationary noise) reduces to the problem of minimum mean square estimation of c_n and repeating $g(\cdot)$ at multiples of T, using the estimated amplitude [35, p. 253], [41].

The above example is closely related to ECG denoising using a data model of the form (5.2). Accordingly, if the inter-beat variations of the ECG were negligible, an ECG would be a *wide-sense cyclostationary* process. In that case, one could *optimally*— in the Wiener filtering sense— filter the ECG as demonstrated in Fig. 5.11: (1) detect the R-peaks, (2) perform synchronous averaging (or *robust weighted averaging* [52]) to find the average ECG beat, and (3) reconstruct the denoised ECG by repeating the average beat at the R-peak locations [41]. Now suppose that $z(t) = x(t) + \eta(t)$ is a signal acquired from a maternal abdominal lead, $x(t)$ is the mECG and we are interested in the background signal $\eta(t)$, which is the fECG plus other noises. In this case, the above algorithm simply reduces to template subtraction: *construct a maternal ECG template and subtract this template by aligning it under the maternal R-peaks of the abdominal leads.* However, since in reality the ECG has RR-interval deviations and morphological variations, instead of simple template subtraction that does not account for beat-wise heart rate and morphological variations, it is better to make the procedure beat-wise adaptive to compensate the beat-wise variations of the ECG (parametrized by γ_n in the data model (5.2)).

For example, the *cardiac phase signal* introduced in Sect. 5.2.2 can be used to compensate the RR-interval deviations by time-warping [83]. The minor beat-wise variations can further be compensated using classical beat alignment techniques [5, 92]. The template subtraction may also be made beat-wise adaptive, using Kalman filtering schemes as detailed in Sect. 5.4.4. In fact, by applying such beat alignment techniques, the beat-wise deviations parametrized by γ_n in (5.2) are compensated and the resulting signal would become cyclostationary. As a result, the optimal cyclostationary Wiener filter for removing the mECG from maternal abdominal recordings is basically a template subtraction in the transformed domain (after compensating the beat-wise deviations of the mECG).

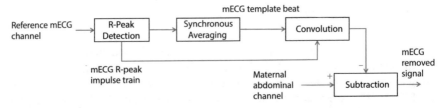

Fig. 5.11 Demonstration of the concept of optimal cyclostationary Wiener filtering for mECG cancellation

5.4.3 Adaptive Filters for fECG Extraction

Adaptive filters are one of the popular filters used for mECG cancellation and fECG extraction. The procedure consists of training an adaptive filter for either removing the mECG using one or several maternal reference channels [67, 104] or directly training the filter for extracting the fetal QRS waves [32, 69]. Ad hoc, adaptive filters such as *partition-based weighted sum filters* [89] and least square error fittings [59] have also been used for this purpose. A comparative study of template subtraction and several adaptive filters including the *least mean squares* (LMS), *recursive least squares* (RLS), and an ad hoc filter coined *echo state neural network* (ESN) was reported in [7, 8].

As demonstrated in Fig. 5.12, adaptive filtering methods for mECG removal either require a reference mECG channel that is morphologically similar to the contaminating waveform or require several channels to approximately reconstruct any morphological shape from the reference channels using adaptive [104], neural networks or neuro-fuzzy inference systems [4]. Both of these approaches are practically inconvenient and have limiting performance since the morphology of the mECG contaminants highly depends on the electrode locations, and it is not always possible to reconstruct the complete mECG morphology from a (linear) combination of the reference electrodes, especially due to the limitations of finite dimensional dipole model of the heart, detailed in Sect. 5.2.1.

5.4.4 Kalman Filters for fECG Extraction

Adaptive methods of mECG cancellation should ideally not rely on the electrode placement and the mECG morphology of the reference channel. This objective has motivated the development of Kalman filters for fECG extraction [63, 64, 74, 82,

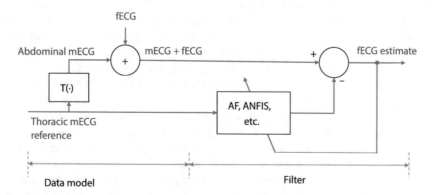

Fig. 5.12 Adaptive filters for maternal ECG cancellation. (Concept adapted from [4])

84]. The Kalman filter and its extensions are adaptive in their nature and are therefore ideal for ECG signals with beat-wise morphological variations.

In [82], an extended Kalman filter (EKF) was suggested for denoising ECG signals recorded from noisy data. The *process model* required for this EKF was based
on an extension of the McSharry-Clifford synthetic ECG model [61, 79]. The EKF
formulation was later used in [74, 84] for removing mECG artifacts from maternal
abdominal recordings. Accordingly, following the volume conduction and dipolar
data models (5.1) and (5.5), we can assume that the maternal abdominal signals
consist of the mECG $s_m(t)$, fECG $s_f(t)$, and background noise $\nu(t)$. For normal ECG,
the mECG and fECG components are pseudo-periodic random processes, which
can be described by a set of dynamic equations. For example, by using the nonlinear
state-space model proposed in [82] for mECG modeling, the following set of process and observation equations can be written for the maternal body surface recorded
signals $x(t)$:

- *Process equations*:

$$\theta(t+1) = \left[\theta(t) + \omega_m(t)\right] \mathrm{mod}(2\pi)$$

$$s_m(t+1) = s_m(t) - \omega_m(t)\sum_k^{i=1} \frac{\alpha_i \theta_i(t)}{b_i^2} \exp\left(\frac{-\theta_i(t)^2}{2b_i^2}\right) + w(t) \tag{5.8}$$

- *Observation equations*:

$$\phi(t) = \theta(t) + v(t)$$
$$x(t) = s_m(t) + s_f(t) + \eta(t) \tag{5.9}$$

where $\theta_i(t) = \left[\theta(t) - \theta_i\right]\mathrm{mod}(2\pi)$ and $\omega_m(t) = 2\pi f_m(t)/f_s$ are the maternal normalized angular velocities, $f_m(t)$ is the instantaneous maternal heart rate in Hertz, f_s
is the sampling frequency in Hertz, α_i, b_i, and θ_i are the amplitude, width, and center
parameters of the ith Gaussian kernel, and k is the number of Gaussian kernels used
for modeling the mECG morphology. In (5.8) and (5.9), $\theta(t)$ and $s_m(t)$ are the state
variables, $\phi(t)$ is the cardiac phase measurement obtained by maternal RR-interval
calculation and a linear phase map as demonstrated in Fig. 5.3, $x(t)$ is the maternal
abdominal ECG measurement, $w(t)$ denotes the process noise, $\nu(t)$ is the phase
measurement noise, and $\eta(t)$ is the ECG measurement noise. According to the procedure detailed in [82], this model can be used in an EKF for estimating the mECG
$\hat{s}_m(t)$. At the same time, the residual signal $x(t) - \hat{s}_m(t)$ (known as the *innovation
process* of the Kalman filter) is an estimate of $s_f(t) + \eta(t)$. The source codes required
for implementing this method— and the other methods detailed in this chapter—
are available online in the *open-source electrophysiological toolbox* (OSET) [76].

An advantage of the Kalman filtering framework is that, besides signal estimation
and denoising, it intrinsically provides confidence intervals for the estimations as well.
By defining $\mathbf{x}(t) = [\theta(t), s(t)]^T$ as the state vector at instant t and $\hat{\mathbf{x}}(t)$ as the *posterior*

estimate of $\mathbf{x}(t)$, the posterior error of the estimation is defined as $\mathbf{e}(t) = \mathbf{x}(t) - \hat{\mathbf{x}}(t)$ with a covariance matrix $\mathbf{P}(t) = \mathbb{E}\left\{\left(\mathbf{e}(t) - \mathbb{E}\{\mathbf{e}(t)\}\right)\left(\mathbf{e}(t) - \mathbb{E}\{\mathbf{e}(t)\}\right)^T\right\}$. The matrix $\mathbf{P}(t)$ is an essential part of all the different variants of the Kalman filter and is calculated and updated as the filter propagates in time. The eigenvalues of this matrix can be used to form an *error likelihood ellipsoid* (also known as *concentration ellipsoid* [100]) that represents the region of highest likelihood for the true state vector $\mathbf{x}(t)$. This likelihood ellipsoid provides a confidence region for the estimated signals.

The overall procedure for removing mECG signals by using the Kalman filtering framework is illustrated in Fig. 5.13 and may be summarized as follows:

1. *Baseline wander removal.* For the reliable extraction of the average mECG templates, the baseline wander of the noisy records should be removed beforehand.
2. *mECG R-peak detection.* These peaks are required for constructing the phase signal $\theta(t)$, which in turn is needed for synchronizing the noisy ECG with the dynamic model in (5.8). They are also used for extracting the mean mECG by synchronous averaging over the maternal heart beats. Depending on the power of the contaminating mECG, as compared with the background signals and noise, the maternal R-peaks may be detectable from the noisy recordings or from an arbitrary chest lead or abdominal channel synchronously recorded with the noisy dataset.
3. *mECG template extraction.* Using the R-peaks, the *ensemble average* (EA) and standard deviation of the mECG are extracted through synchronous averaging. Several methods have been proposed in the literature for synchronous averaging. One of the most effective approaches is the *robust weighted averaging* method [51], which outperforms conventional EA extraction methods and is useful for noisy nonstationary mixtures.
4. *Model fitting.* As proposed in [20, 82], by using a nonlinear least square estimation, the parameters of the Gaussian kernel defined in (5.8) are found, such that the model will best fit the mean mECG waveform.
5. *Covariance matrix calculations.* The standard deviation of the average mECG is used to find the entries of the process and observation noise covariance matrices, as required for (extended) Kalman filtering.

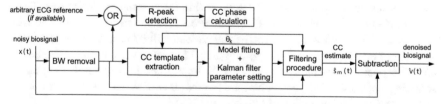

Fig. 5.13 The overall denoising scheme. As shown in this figure, the R-peaks of the contaminating signals (CC) may be detected either from an arbitrary reference ECG or from the noisy biosignal after baseline wander (BW) removal. (Adapted from [84])

6. *Filtering*. Having the required model parameters, the mECG may be estimated by the EKF framework and the desired background signal (fECG plus noise) is found from $\hat{v}(t) = x(t) - \hat{s}_m(t)$.

7. *fECG post-processing*. The residual signals containing fECG and noise are post-processed for improving the fECG signal quality. Various methods such as an adaptive filter, a wavelet denoiser, or even a secondary EKF stage (this time customized for fECG denoising) can be used in this stage.

Note further that for online applications or denoising long nonstationary datasets, all the dynamic model parameters and the covariance matrices can be updated over time, by recalculating them from the most recent cardiac beats. Further details regarding the Kalman filter–based approach and its extensions such as the extended Kalman smoother (EKS), unscented Kalman filter (UKF), and H-infinity filter can be followed from [42, 63, 82, 84].

In Fig. 5.14a, the first channel of the DaISy fECG dataset is used for illustration [29]. The mECG estimate and the fetal residual components are depicted in Fig. 5.14b, c. As a post-processing step, the extended Kalman filtering algorithm is applied to the residual fetal components, this time by training the filter parameters over the fECG. The post-processed fECG are depicted in Fig. 5.14d. From these results, it is seen that the Kalman filter is very effective for the extraction of fECG components from noisy maternal abdominal mixtures, even from as few as a single channel. However, as noticed from Fig. 5.14d, between $t = 6$ s and $t = 7$ s, the filter has failed to discriminate between the maternal and fetal components when the ECG waves of the mother and fetus have fully overlapped in time. The reason is that when the maternal and fetal components coincide in time, there are no other a priori information for separating the maternal and fetal components. This is in fact an intrinsic limitation of single-channel methods, which motivates the application of multichannel recordings.

As noted before, an important feature of Bayesian filtering is the ability of predicting the accuracy of the estimates. For the Kalman filter, this is readily achieved through the calculation of the error covariance matrix $\mathbf{P}(t)$. Suppose that the entry of the covariance matrix $\mathbf{P}(t)$ corresponding to the ECG estimate is denoted as $\sigma(t)^2$ and the ECG estimation error is Gaussian, then the estimated ECG is bounded within the $\pm\sigma(t)$ envelope in 68% of the sample points. This is due to the fact that approximately 68% of the values drawn from a Gaussian distribution are within one standard deviation away from the mean, about 95% of its values are within two standard deviations, and about 99.7% lie within three standard deviations. These probabilities are different for non-Gaussian errors obtained by a nonlinear estimator such as the EKF. However, the $\pm\sigma(t)$ envelope can still be used as an approximate measure of error spread [100, p. 79]. In Fig. 5.15, several beats of the fECG before and after post-processing by an extended Kalman filter, together with their corresponding $\pm\sigma(t)$ and $\pm3\sigma(t)$ envelopes, are plotted. It is seen that the error envelopes provide the confidence region of the denoised fECG.

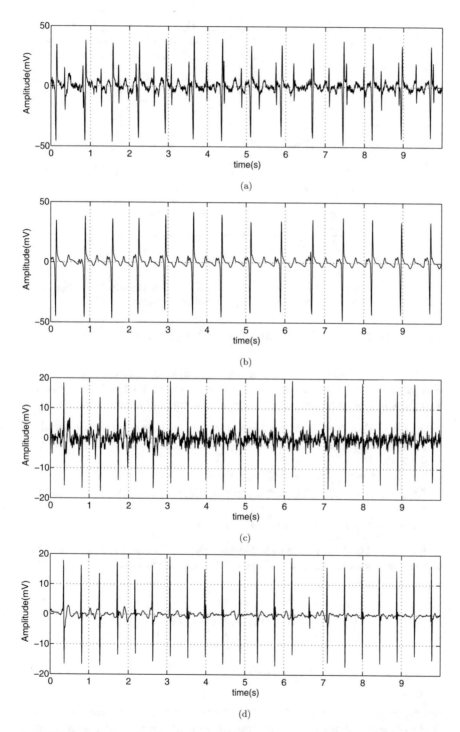

Fig. 5.14 The first channel of the DaISy dataset [29], recorded from a maternal abdominal lead before and after the EKF procedure. (Adapted from [74]). (**a**) Original. (**b**) EKF of the maternal ECG. (**c**) Residual fetal signal. (**d**) Fetal signal after post-processing

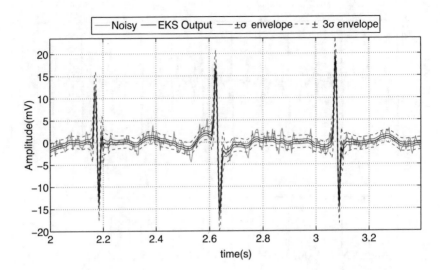

Fig. 5.15 Several fetal ECG beats adapted from Fig. 5.14, before and after the post-processing EKF, together with the $\pm\sigma(t)$ and $\pm 3\sigma(t)$ confidence envelopes. (Adapted from [74])

5.5 Multichannel Fetal Electrocardiogram Extraction

Due to the limitations of single-channel fECG analysis detailed in the previous section, advanced fECG extraction algorithms are commonly multichannel. Some of the advantages of multichannel fECG acquisition and analysis are as follows:

- Improved SNR due to spatial filtering and joint analysis of multiple channels
- Robustness to fetal position and displacement due to the spatial diversity of the leads
- Robustness to the possible detachment of a few of the electrodes
- Ability to extract the fECG even during overlapping of ECG waves of the mother and fetus
- Obtaining multiple perspectives of the fetal heart

Reconsidering the maternal abdominal recordings data model (5.1), in the multichannel case, it can be represented in the following matrix form:

$$
\mathbf{x}(t) = \begin{bmatrix} \mathbf{H}_m & \mathbf{H}_f & \mathbf{H}_v \end{bmatrix} \begin{bmatrix} \mathbf{s}_m(t) \\ \mathbf{s}_f(t) \\ \mathbf{v}(t) \end{bmatrix} + \mathbf{n}(t) = \mathbf{As}(t) + \mathbf{n}(t)
$$

$$
= \sum_{p}^{k=1} \mathbf{a}_k s_k(t) + \mathbf{n}(t)
$$

(5.10)

where $p \triangleq m + l + k$ is the total effective number of sources due to the maternal and fetal ECG and structured noises, $\mathbf{A} = [\mathbf{a}_1, \ldots, \mathbf{a}_p] \in \mathbb{R}^{n \times p}$ is the overall source-sensor mixing matrix (or the lead-field matrix), and $\mathbf{s}(t) = [s_1(t), \ldots, s_p(t)] \in \mathbb{R}^p$ contains all the cardiac sources and structured noise components.

The objective of multichannel analysis is to recover an estimate of $\mathbf{s}(t)$ (or more specifically $\mathbf{s}_f(t)$) from $\mathbf{x}(t)$, using the available assumptions regarding the mECG, fECG, and noises. A classical approach to solving this problem is to estimate the matrix $\mathbf{B} \in \mathbb{R}^{p \times n}$, such that $\mathbf{BA} = \mathbf{I}$. Therefore,

$$\mathbf{y}(t) = \mathbf{Bx}(t) = \mathbf{s}(t) + \mathbf{Bn}(t) \tag{5.11}$$

which is a noisy estimate of the source vector $\mathbf{s}(t)$. Since both the source vector $\mathbf{s}(t)$ and the mixing matrix $\mathbf{A}$ are unknown, the problem is categorized as a *blind or semi-blind source separation* (BSS) problem [26]. In this problem, if the number of observed channels is equal to or greater than the effective number of sources, i.e., $n \geq p$, and $\mathbf{A}$ is non-singular, then the observed mixture is *determined* or *over-determined*. Therefore, noting that $\mathbf{s}_m(t)$, $\mathbf{s}_f(t)$, and $\mathbf{v}(t)$ can be considered as groups of statistically independent sources with inter-independence and intra-dependencies, BSS algorithms such as (noisy) independent component analysis (ICA) [13, 28, 108], semi-blind source separation algorithms such as periodic component analysis (πCA) [83], and more recently nonstationary component analysis (NSCA) [42] have been effectively used to solve this problem. The general challenges of this problem are as follows:

1. *Amplitude and sign ambiguity*: An intrinsic ambiguity of the multichannel data model (5.10) is that the source vector amplitude and sign may not be retrieved merely from the measurements $\mathbf{x}(t)$. This can be explained by the fact that exchanging an arbitrary non-zero scaling factor α and $1/\alpha$ between the kth column of the matrix $\mathbf{A}$ and the source $s_k(t)$ does not change the measurements. Therefore, there is no way to retrieve the source amplitudes and sign from the measurements alone.

2. *Estimated source order:* Retrieving the order of sources is another limitation that may not be resolved from the measurements alone (without other priors or constraints). The reason is that taking an arbitrary permutation matrix $\mathbf{P}$, $\mathbf{As}(t)$, and $\mathbf{APP}^T\mathbf{s}(t)$ is identical.

3. *Noisy mixtures:* It is clear from the right-hand side of (5.11) that even if the separation matrix $\mathbf{B}$ is perfectly estimated, i.e., $\mathbf{BA} = \mathbf{I}$, due to the noise term $\mathbf{Bn}(t)$, then the resulting mixture can remain noisy, except for the non-probable special case that the observation noise lies in the *null space* of the separation matrix $\mathbf{B}$, resulting in $\mathbf{Bn}(t) = \mathbf{0}$. Otherwise, the noise can even be amplified and the desired components, such as the fECG, may in cases be totally obscured by noise. In fact, the problem due to full-rank observation noise is twofold. On the one hand, the noise hampers the estimation of the separation matrix. On the other hand, it remains or is even amplified during source separation. Therefore, whenever possible, it is better to minimize or remove the channel-wise full-rank noise before source separation. In the latter case (channel-wise noise removal), any processing of the multichannel data should be performed by using filters that

approximately have a *linear phase* (constant *group delay*) over the bandwidth of interest. Moreover, the difference between the group delays of the filters applied to different channels should be negligible, as compared with the sampling time of the data, to avoid the displacement of the components of different channels during preprocessing. This is a fundamental requirement for synchronous multi-channel analysis, which has been underemphasized in the literature.

4. *Non-punctual sources:* The heart is not a punctual source. This fact has several implications on fECG extraction, including the following: (1) the fECG mor-phology can change as the fetus moves with respect to the maternal body surface leads; (2) during source separation, depending on the heart-sensor distance and the SNR of the measurements, more than one source is associated with the mother and the fetus. The notion of effective number of sources detailed in Sect. 5.2.1 corresponds to this fact. It has been previously shown that even though, among the extracted sources, only a few might visually resemble the fECG, when one applies synchronous averaging to the different channels extracted by BSS algorithms (by aligning the R-peak positions and averaging over several beats), the fECG emerges from all channels. This point was first illustrated in [80] and justified in [74] using multi-pole expansion of body surface potentials. An example of adult and fetal ECG obtained by synchronous averaging after applying a typical ICA algorithm is shown in Fig. 5.16. This implies that for non-punctual sources, perfect separation of the sources (maternal and fetal ECG) is not fully achieved. However, in practice, the number of cardiac source signals extracted from multichannel ECG— including maternal abdominal recordings—are limited by the number of channels, distance to the heart, and the SNR of the recordings.

5. *Low-rank measurements:* If the number of abdominal channels are insufficient ($n < p$) or when the maternal-fetal mixture is singular (e.g., due to the closeness of the sensors or special fetal positioning in the womb), the signal mixture is

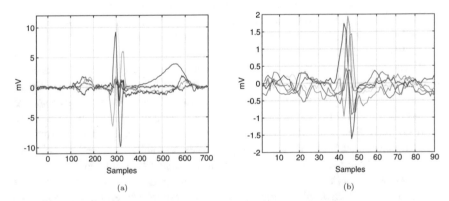

(a)   (b)

Fig. 5.16 An illustration of the concept of non-punctuality of the cardiac sources resulting in multidimensionality of the components extracted from adult and fetal ECG. Synchronous averag-ing has been performed over the different channels extracted by independent component analysis to demonstrate the existence of the ECG components in all channels. (**a**) Adult ECG. (**b**) Fetal ECG

under-determined. In this case, due to the rank-deficiency of the mixture, linear transforms are unable to separate the maternal and fetal subspaces [41, 86].

6. *Time-variant mixtures:* When the mixing matrices $\mathbf{H}_m$ and $\mathbf{H}_f$ are functions of time, e.g., due to fetal movement during signal acquisition, the sources may no longer be retrieved by stationary source separation algorithms. In this context, adaptive source separation algorithms are required [16]. These methods have also been specifically used for online fECG extraction [33, 41].

In the sequel, some of the different approaches of fECG extraction from multi-channel recordings are reviewed.

5.5.1 Independent Component Analysis

Independent component analysis (ICA) is the most common class of algorithms for solving blind and semi-blind source separation (BSS) problems such as (5.10), where both the mixing matrix $\mathbf{A}$ and the source vector $\mathbf{s}(t)$ are unknown (with or without noise) [26]. The problem of retrieving the sources and mixing matrix at the same time is clearly ill-posed. Therefore, additional assumptions and priors about the source and/or mixture are required. In ICA, one seeks linear mixtures of the form $\mathbf{y}(t) = \mathbf{B}\mathbf{x}(t)$, which maximize some measure of statistical independence between the estimated sources, also known as a *contrast function*.

Many ICA algorithms attempt to solve the problem in several phases, for example by first pre-whitening and sphering the data by principal component analysis (PCA) (Fig. 5.17). Pre-whitening acts as an intermediate step for achieving independence and only leaves the estimation of a rotation matrix to achieve independence.

An algebraic approach to ICA is to seek the separation matrix $\mathbf{B}$, such that it diagonalizes a set of matrices containing second- or higher-order statistics derived from the multichannel recordings [26]. For signals with temporal structure, there are various algorithms that use this algebraic approach. Considering that no more than two matrices can be simultaneously diagonalized by using a single linear transform, many algebraic algorithms have been developed for the approximate joint diagonalization of such matrices. The first and most widely used algorithm in this context is known as *joint approximate diagonalization of eigenmatrices* (JADE) [15, 17]. To date, fECG extraction has been one of the classical biomedical applications for testing and comparing various ICA algorithms. Some of the pioneer contributions in this area include the following studies: [13, 28, 108].

$$\mathbf{x}(t) \xrightarrow[\text{(rotation)}]{\text{Whitening}} \mathbf{v}(t) \xrightarrow[\text{(scaling)}]{\text{Sphering}} \mathbf{z}(t) \xrightarrow[\text{(rotation)}]{\text{ICA}} \mathbf{y}(t)$$

Fig. 5.17 General scheme of ICA algorithms with spatial pre-whitening

5.5.2 Independent Subspace Analysis

Independent subspace analysis (ISA) has been introduced as a variant of ICA for problems in which one deals with groups of signals having inter-group indepen-dence and intra-group dependencies. ISA was first introduced in [25] and mathe-matically developed in [13], where the notion of ICA was generalized to the notion of *multidimensional ICA*. Accordingly, ISA relies on the idea of *vector-valued* com-ponents rather than *scalar* source signals. The first—and most commonly studied—application of ISA has been for fECG extraction. Throughout the chapter, we have learned that the cardiac signals of either the mother or the fetus are generally multi-dimensional. Therefore, the maternal and fetal ECG components form signal sub-spaces with internal dependencies, while the components of the maternal and fetal subspaces are independent from each other.

ISA may be realized by applying an initial ICA step on mutichannel observations and then empirically regrouping the independent components that belong to the same subspace from prior knowledge of the subspace structures to achieve a *canoni-cal representation* of each subspace. In fact, there is an intrinsic ambiguity in retrieving the components inside the subspaces, which may not be resolved with the same measure of independence used for subspace separation. In other words, from the source separation viewpoint, no representation of the extracted mECG and fECG components inside their signal subspaces can be considered to be better than the other. Therefore, the components that belong to the same subspace are regrouped after the initial ICA step. However, the challenges of ISA are as follows:

1. Finding the dimensions of each subspace [13]
2. Automatic regrouping of the components [6, 95, 103]
3. Studying the impact of subspace distances and noise on the stability of the extracted subspaces [37, 62].

For fECG extraction, previous studies have focused on the feasibility of extracting the independent subspaces [13, 28] and regrouping strategies [6].

5.5.3 Generalized Eigenvalue Decomposition

Although ICA and ISA are very effective for fECG extraction, they do not make explicit use of the pseudo-periodicity of the maternal and fetal ECG and the fact that multiple sources may correspond to the mECG and the fECG (due to the non-punctuality of the cardiac sources detailed before). In order to be used in fully auto-mated algorithms, it is also convenient to be able to rank the extracted sources corresponding to the mECG and/or fECG automatically. These requirements resulted in the development of source separation algorithms, which are specifically customized for cardiac signals. Algorithms such as periodic component analysis (πCA) [83] and nonstationary component analysis [42] were developed for this purpose. These methods are based on an algebraic transform known as generalized eigenvalue decomposition, which was previously used in one of the basic source separation algorithms known as AMUSE [99].

For real symmetric matrices $\mathbf{A}, \mathbf{B} \in \mathbb{R}^{n \times n}$, *generalized eigenvalue decomposition* (GEVD) of the matrix pair $(\mathbf{A}, \mathbf{B})$ consists of finding $\mathbf{W} \in \mathbb{R}^{n \times n}$ and $\Lambda \in \mathbb{R}^{n \times n}$, such that

$$\mathbf{W}^T \mathbf{A} \mathbf{W} = \Lambda$$
$$\mathbf{W}^T \mathbf{B} \mathbf{W} = \mathbf{I}_n$$

(5.12)

where $\Lambda = \mathbf{diag}\,(\lambda_1, \ldots, \lambda_n)$ contains the generalized eigenvalues corresponding to the eigenmatrix $\mathbf{W} = [w_1, \ldots, w_n]$, with real eigenvalues sorted in descending order on its diagonal. Symmetric positive definite matrix pairs have real positive eigenvalues and the first eigenvector $\mathbf{w} = \mathbf{w}_1$ maximizes the *Rayleigh quotient* [96]:

$$J(\mathbf{w}) = \frac{\mathbf{w}^T \mathbf{A} \mathbf{w}}{\mathbf{w}^T \mathbf{B} \mathbf{w}}$$

(5.13)

It can be shown that all ICA methods based on pre-whitening can be eventually converted into a GEVD problem of two (problem-specific) matrices [83]. Therefore, in semi-blind source separation problems, in which prior knowledge regarding the underlying components exists, the problem of source separation can be considered as a *matrix design problem*. The performance of GEVD-based source separation and generic methods for choosing the proper matrix pair have been addressed in previous research works [105, 107].

GEVD can, for example, be used for the separation of temporally correlated (or periodic) sources from other signals. For example, for a zero-mean wide-sense stationary or cyclostationary real observation vector $\mathbf{x}(t)$, the covariance matrix is:

$$\mathbf{C}_x(\tau) = \mathbb{E}_t \left\{ \mathbf{x}(t+\tau)\mathbf{x}(t)^T \right\}$$

(5.14)

where $\mathbb{E}_t\{\cdot\}$ indicates averaging over t. The AMUSE algorithm is a source separation algorithm that jointly whitens the data and diagonalizes $\mathbf{C}_x(\tau)$ for some arbitrary τ, i.e., the solution of the GEVD problem of the matrix pair $(\mathbf{C}_x(\tau), \mathbf{C}_x(0))$ [70, 99]. What hampers the performance of GEVD for source separation is the fact that real-world sources are rarely fully periodic. Therefore, more advanced source separation algorithms use (approximate) joint diagonalization of more than two matrices, which are more robust to data outliers and computational errors as compared with AMUSE [10, 14]. In this context, the second-order blind identification (SOBI) algorithm is an example of a time-domain algorithm that whitens the data and approximately diagonalizes $\mathbf{C}_x(\tau)$ for several time-lags τ [10]. Similar time-domain methods have also been proposed for cyclostationary sources, in which the data is again pre-whitened and matrices corresponding to cyclostationary statistics of the dataset are (approximately) diagonalized [34]. An alternative approach is to use signal priors such as the pseudo-periodicity and "bumpy" shape of the ECG, as detailed below.

5.5.4 Periodic Component Analysis

In (pseudo-)periodic component analysis (πCA),[1] the matrix pair $(\mathbf{C}_1, \mathbf{C}_0)$ is jointly diagonalized by GEVD, where $\mathbf{C}_0 = \mathbf{C}_x$ is the covariance matrix of $\mathbf{x}(t)$ and $\mathbf{C}_1$ is the variable-period version of the lagged-covariance matrix (5.14), using the time-varying period of the ECG defined in (5.4):

$$\mathbf{C}_1 = \mathbb{E}_t \left\{ \mathbf{x}(t+\tau_t)\mathbf{x}(t)^T \right\} \tag{5.15}$$

In order to assure the symmetry of $\mathbf{C}_1$ and the realness of its eigenvalues, the following step is applied before GEVD:

$$\mathbf{C}_1 \leftarrow \frac{\left(\mathbf{C}_1 + \mathbf{C}_1^T\right)}{2} \tag{5.16}$$

Next, considering $\mathbf{W}$ as the joint diagonalizer of the matrix pair $(\mathbf{C}_1, \mathbf{C}_0)$, the linear transform

$$\mathbf{y}(t) = \mathbf{W}^T \mathbf{x}(t) \tag{5.17}$$

extracts uncorrelated sources $\mathbf{y}(t) = [y_1(t), \ldots, y_n(t)]^T$ with maximal correlation at time-variant periods τ_t, which is the heart rate of interest. Therefore, $\mathbf{y}(t)$ ranks the sources in order of similarity with the desired heart rate. In other words, $y_1(t)$ is the most periodic component and $y_n(t)$ is the least periodic with respect to the R-peaks of the ECG. This method is flexible in the cardiac period used for source separation. For instance, for fECG extraction, let $\theta_m(t)$ and $\theta_f(t)$ be the maternal and fetal ECG phases found from the maternal and fetal R-peaks (as defined in Sect. 5.2.2.2) and $\mathbf{C}_m$ and $\mathbf{C}_f$ represent the lagged covariance matrices of the maternal and fetal heart rates found by averaging (5.15) over the maternal and fetal periods τ_t^m and τ_t^f, respectively. Then different variants of GEVD are obtained if the matrix $\mathbf{C}_1$ used in GEVD is set to any of the following matrices [83]:

$$\left(\mathbf{C}_1, \mathbf{C}_0\right) = \left(\mathbf{C}_m, \mathbf{C}_x\right) \tag{5.18a}$$

$$\left(\mathbf{C}_1, \mathbf{C}_0\right) = \left(\mathbf{C}_f, \mathbf{C}_x\right) \tag{5.18b}$$

---

[1] The term πCA was originally coined in [87], for extracting periodic signals, which resulted in GEVD of a pair matrices as in AMUSE [99].

$$\left(\mathbf{C}_1, \mathbf{C}_0\right) = \left(\mathbf{C}_m, \mathbf{C}_f\right) \tag{5.18c}$$

If we assume the data to be pre-whitened, the diagonalization of the matrices defined in (5.18) is respectively equivalent to finding (a) the most periodic components with respect to the mECG, (b) the most periodic components with respect to the fECG, and (c) the most periodic components with respect to the mECG while being the least periodic components with respect to the fECG. In this latter case, the extracted components should gradually change from the mECG to the fECG, from the first to the last component, but the components are not necessarily uncorrelated. It should of course be noted that the last two cases are difficult to implement in practice, as they require the prior extraction of the fetal R-peaks to form the $\mathbf{C}_f$ matrix. Another reservation is for abnormal maternal cardiac signals, for which the signals are no longer regular or pseudo-periodic and a measure of pseudo-periodicity can fail for mECG and fECG source separation.

5.5.5 Nonstationary Component Analysis

The reservations regarding possible abnormal mECG and the difficulty of fECG R-peak identification in noise have motivated source separation algorithms that are merely based on rather regular spiky or bumpy shapes of the maternal and fetal ECG. The theory is based on source separation algorithms for variance-nonstationary source mixtures, which is a special case of methods known as *nonstationary component analysis* (NSCA) [42, 71, 106]. Accordingly, let us consider multivariate signals $\mathbf{x}(t) \in \mathbb{R}^n$ ($t \in \mathcal{T}$), where $\mathcal{T}$ denotes the set of available discrete-time samples and $\mathcal{P} \subset \mathcal{T}$ is a subset of these samples, which are considered as being *nonstationary* or *odd events* that do not follow the (average) background model in certain aspects. For our application, they can correspond to the maternal or fetal QRS complexes. In this case, a sample-wise *hypothesis test* can be performed for the identification of the temporally nonstationary events:

$$\begin{aligned}
\mathcal{H}_0 &: t \notin \mathcal{P} \\
\mathcal{H}_1 &: t \in \mathcal{P}
\end{aligned} \tag{5.19}$$

Denoting the subset of samples that satisfy the alternative hypothesis H_1 with $\mathcal{P}$, a special case of GEVD is obtained by finding the matrix $\mathbf{W}$, which satisfies (5.12) for $\mathbf{A} = \mathbb{E}_u \left\{ \mathbf{x}(u) \mathbf{x}(u)^T \right\}$ and $\mathbf{B} = \mathbb{E}_t \left\{ \mathbf{x}(t) \mathbf{x}(t)^T \right\}$, where $\mathbb{E}_t \{\cdot\}$ and $\mathbb{E}_u \{\cdot\}$ denote averaging over all time samples $t \in \mathcal{T}$ and $u \in \mathcal{P}$, respectively. Using this matrix, the linear transform $\mathbf{y}(t) = \mathbf{W}^T \mathbf{x}(t)$ extracts n uncorrelated channels with maximal energy over the subset of time samples $u \in \mathcal{P}$. Applying this method for ECG extraction, $\mathbf{W}$ retrieves uncorrelated linear mixtures of $\mathbf{x}(t)$ with maximal energy during the QRS complex.

As detailed in [42], in the simplest case, the nonstationary sample set $\mathcal{P}$ can be identified by thresholding the time-varying power of an arbitrary reference channel $r(t)$ (which can even be one of the channels of $\mathbf{x}(t)$, or a mixture of them) over a sliding window of length w:

$$P_w(t) = \frac{1}{w} \sum_{a=-\frac{w}{2}}^{\frac{w}{2}} |r(t-a)|^2 \tag{5.20}$$

The ratio of $P_w(t)$ for two windows of lengths $w = w_1$ and $w = w_2$ ($w_2 \gg w_1$) can be used as a measure for detecting fast local nonstationary epochs within a slowly varying (or stationary) background activity:

$$\rho(t) = \frac{P_{w_1}(t)}{P_{w_2}(t)} \tag{5.21}$$

which is the *local power envelope* (LOPE) of the reference channel. For a global measure, the denominator $P_{w_2}(t)$ can be replaced with the average signal power P_∞. The values of $\rho(t)$ significantly smaller or larger than 1 correspond to time epochs that are different (nonstationary) from the background activity. The rationale behind the above definition is that a stationary signal, such as the non-ECG background signals and noises, has a consistent energy profile over time, and notable deviations of the LOPE from unity (with appropriate window lengths w_1 and w_2) are indicators of nonstationary epochs such as the maternal and fetal QRS complexes. Therefore, the LOPE can be used to extract the time epochs of the maternal or fetal QRS as follows:

$$\theta_{\text{LPE}} = \{t \mid \rho(t) \ge \zeta_u \text{ or } \rho(t) \le \zeta_l, t \in T\} \tag{5.22}$$

where ζ_u and ζ_l are predefined upper and lower thresholds satisfying $\zeta_u > 1 > \zeta_l \ge 0$. In [42], other indexes based on the *innovation process* of an extended Kalman filter trained over the mECG were proposed for the identification and extraction of the fECG.

5.5.6 Approximate Joint Diagonalization Using ECG-Specific Priors

Maternal and fetal ECG source separation from background noise can benefit from the advantages of methods such as πCA and NSCA at the same time. Suppose that the matrices $\mathbf{C}_i$ ($i = 1, \ldots, K$) are positive semi-definite matrices containing second- or higher-order statistics regarding the maternal and fetal ECG. For example, the matrices can be the lagged-covariance matrices corresponding to the maternal or fetal heart, or the covariance matrices obtained by energy thresholding, as in NSCA. We may now seek the joint approximate diagonalizer $\mathbf{W} \in \mathbb{R}^{n \times n}$, such that the matrices

$$\mathbf{W}^T \mathbf{C}_i \mathbf{W} = \Lambda_i, \quad i = 1, \ldots, K \tag{5.23}$$

are "as diagonal as possible." It is known that for $K > 2$, the diagonalization is only achieved approximately by using different variants of approximate joint diagonalization (AJD). Depending on the application and diagonalization algorithm, in order to achieve uncorrelated sources, the total covariance matrix $\mathbf{C}_x = \mathbb{E}\left\{\left(\mathbf{x}(t) - \mathbf{m}_x\right)\left(\mathbf{x}(t) - \mathbf{m}_x\right)^T\right\}$ ($\mathbf{m}_x = \mathbb{E}\left\{\mathbf{x}(t)\right\}$) may also be among the set of matrices to be diagonalized.[2] The approach based on AJD is more robust as compared with πCA and NSCA, which only work with two matrices. It is also more effective than JADE and other generic ICA algorithms, as it uses specific features of the ECG of the mother and the fetus. However, the order of sources is no longer guaranteed in AJD.

5.5.7 Illustration

The DaISy fECG dataset is used for illustration [29]. This sample data consists of five abdominal and three thoracic channels recorded from the abdomen and chest of a pregnant woman at a sampling rate of 250 Hz. The eight channels of the dataset are depicted in Fig. 5.18.

The result of applying independent subspace decomposition [13], using the JADE algorithm [15, 17], is depicted in Fig. 5.19. Accordingly, the first, second, third, and fifth components correspond to the mECG subspace, the fourth and eighth components correspond to the fECG, and the sixth and seventh components are noise.

By performing R-wave detection on the last maternal thoracic channels of Fig. 5.18 (channel eight), the mECG phase $\theta_m(t)$ is calculated as detailed in Sect. 5.2.2.2. Next, the time-varying mECG period τ_t^m is calculated, from which the matrix $\mathbf{C}_m$ and the generalized eigenmatrix $\mathbf{W}$ (the joint diagonalizer) of the ($\mathbf{C}_m$, $\mathbf{C}_x$) pair are found and their columns are sorted in descending order of the corresponding eigenvalues. The resultant periodic components calculated from (5.17) are depicted in Fig. 5.20. Accordingly, the first component, which corresponds to the largest eigenvalue, has the most resemblance with the mECG, while as the eigenvalues decrease, the signals become less similar to the mECG. Although two of the

---

[2] Enforcing the diagonalization of $\mathbf{C}_x$ guarantees decorrelation of the extracted sources at a cost of consuming $n(n - 1)/2$ degrees of freedom of the matrix $\mathbf{W}$. This is why some BSS algorithms do not enforce whitening or sphering but rather include the covariance matrix among the approximately diagonalized set of matrices at a cost of reduced performance [49].

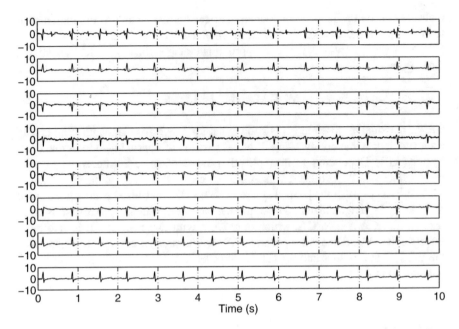

Fig. 5.18 The DaISy dataset consisting of five maternal abdominal and three thoracic channels [29]

extracted components (components six and seven) are the fetal components, the extraction of the fECG has not been explicitly enforced by the algorithm. This can be explained by considering that πCA is ranking the extracted components according to their resemblance with the mECG period, while the fetal components do not resemble the mECG when they are averaged synchronously with respect to the maternal R-peaks. Therefore, the order of appearance of the fECG among the extracted components is unprecedented, merely as components that are uncorrelated with the mECG and the other signals.

As explained in Sect. 5.5.4, it is also possible to consider the fECG periodicity in the matrix $\mathbf{C}_f$, which requires the fetal R-peaks for extracting the time-varying fetal period τ_t^f. To illustrate this case, the fECG component extracted by JADE in the fourth channel of Fig. 5.19 is used for fetal R-peak detection and phase calculation. Having calculated the fECG phase $\theta_f(t)$, GEVD is applied to $(\mathbf{C}_f, \mathbf{C}_x)$ to extract the periodic components of the fECG. The resultant periodic components are depicted in Fig. 5.21. In this case, it is observed that the extracted components are ranked according to their resemblance with the fECG.

The next results correspond to the last type of covariance matrix defined in (5.18) by performing GEVD over the matrix pair $(\mathbf{C}_m, \mathbf{C}_f)$ and calculating the periodic components from (5.17). The resulting components are depicted in Fig. 5.22. As

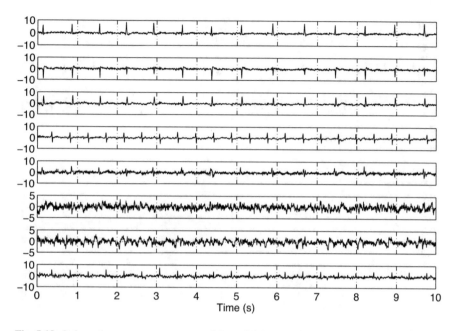

Fig. 5.19 Independent components extracted from the dataset of Fig. 5.18, using the JADE algorithm. Notice that components 1, 2, 3, and 5 correspond to the maternal subspace and components 4 and 8 to the fetal subspace

expected, the first component mostly resembles the mECG, the last component the fECG, and the intermediate components are blended from the maternal to fetal ECG plus noise. Note that in this case, the extracted components are no longer uncorrelated, since the covariance matrix of the data has not been diagonalized.

The next illustration corresponds to the NSCA algorithm. In this case, the local power envelope index detailed in Sect. 5.5.5 is used to detect the local power envelope from the first channel of Fig. 5.18. Considering a typical fetal QRS length of approximately 50 ms, the sliding window lengths of the nonstationarity detector in (5.21) were set to $w_1 = 10$ ms and $w_2 = 200$ ms. The local power envelopes detected by these window lengths can belong to either the mECG or fECG. Therefore, the local peak envelopes of the mECG were independently detected from the last maternal thoracic channel (as a channel which does not have any dominant fetal R-peak due to the electrode location). For this channel, the sliding window lengths were set to $w_1 = 20$ ms and $w_2 = 400$ ms, which are adapted for detecting the mECG segments by thresholding. Next, according to the fusion technique explained in [42], the temporally nonstationary epochs of channel one were excluded from the nonstationary epochs of channel eight, resulting in time instants, which mainly correspond to the fECG and not the mECG. The resulting nonstationary time epochs were used to calculate the required NSCA covariance matrix according to the hypothesis test

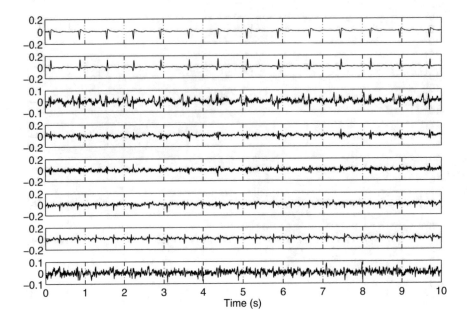

Fig. 5.20 Periodic components extracted by πCA, from the dataset of Fig. 5.18, with *maternal* ECG beat synchronization. The maternal ECG contribution has reduced from top to bottom

(5.19). Finally, GEVD was performed on the covariance matrices and the sources were obtained from (5.17). The results of this method together with the detected nonstationary time epochs are shown in Fig. 5.23, where it is observed that the fECG is successfully extracted and the components are ranked from top to bottom according to their similarity to the fECG. Furthermore, it is seen that the method has been able to extract the fECG even during the temporal overlaps of the mECG and fECG, despite the fact that some of the fetal QRS peaks have not been considered among the temporally nonstationary epochs (notice the missed fetal R-peaks at $t = 1.0, 1.8, 4.0,$ and 4.8 seconds in the nonstationary epochs of Fig. 5.23a). Further details regarding this example can be found in [42].

5.6   Advanced Methods for Fetal ECG Extraction

In this section, some of the advanced methods, which have been developed in the literature for fECG extraction under special circumstances, such as low-rank and time-variant mixtures are reviewed.

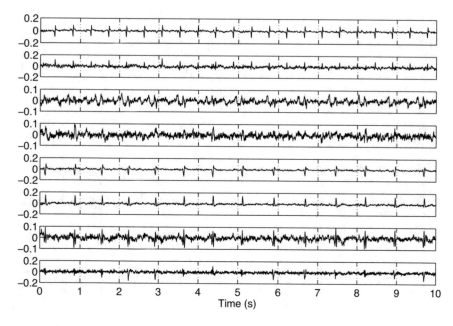

Fig. 5.21 Periodic components extracted by πCA, from the dataset of Fig. 5.18, with *fetal* ECG beat synchronization. It is observed that the fetal ECG contribution reduces from top to bottom

5.6.1 Low-Rank Measurements and Nonlinearly Separable Fetal and Maternal ECG

As noted throughout the chapter, due to the number and placement of the electrodes, and also the fetal positioning, the maternal abdominal recordings can become rank deficient. As a result, in some cases, the fetal and maternal ECG may remain inseparable using any of the aforementioned linear transforms. In these cases, nonlinear methods can be used to separate the maternal and fetal subspaces, or additional synthetic channels can be added to compensate the rank deficiency of the mixtures.

In order to solve the non-separability of the mECG, it has been proposed to synthetically generate q excess "clean" mECG channels—i.e., synthetic channels that resemble the mECG, but do not have any fECG—and to augment the excess channels as auxiliary channel(s) $\mathbf{x}_a(t) \in \mathbb{R}^q$ with the original measured signals [41]:

$$\bar{\mathbf{x}}(t) = \begin{bmatrix} \mathbf{x}(t) \\ \mathbf{x}_a(t) \end{bmatrix} \tag{5.24}$$

where $\bar{\mathbf{x}}(t) \in \mathbb{R}^{n+q}$. It was shown in [41] that the q additional synthetic channels amend the rank deficiency of the problem and help in obtaining a determined or over-determined mixture from which the fECG could be extracted using conventional ICA, πCA, or NSCA algorithms. Apparently, the auxiliary channel

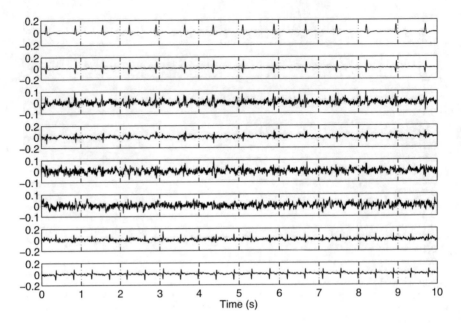

Fig. 5.22 Periodic components extracted by πCA from the dataset of Fig. 5.18, with *maternal and fetal* ECG beat synchronization. The maternal (fetal) ECG contribution reduces (increases) from top to bottom

generation and augmentation is a nonlinear procedure, which utilizes the maternal signals' null space. To implement this method, a channel that resembles the maternal abdominal leads, but is not exactly the same as the other abdominal recorded channels, is needed, which at the same time prevents the multichannel data from becoming singular and does not contain any traces of the fECG.

The ECG cyclostationarity detailed in Sect. 5.4.2, together with the realistic ECG generator described in 5.2.2, provides the means of constructing the required synthetic maternal abdominal ECG. For this, a set of reference channels are selected. Next, the average mECG morphology is calculated by weighted averaging [52]. Either the average morphology is repeated directly at the positions of the maternal R-peaks to construct a synthetic auxiliary channel (according to Sect. 5.2.2) or the mECG is extracted by single-channel adaptive or extended Kalman filtering, as detailed in Sects. 5.4.3 and 5.4.4. The resulting mECG channels are next augmented with the original channels according to (5.24). The augmented data is finally given to multichannel source separation algorithms to recover the maternal and fetal ECG components. Note that this technique may not generally be proved to resolve the problem of rank deficiency, as it is data dependent. However, as demonstrated in [41], it has been shown to resolve the rank deficiency of some of the most popular fECG datasets available online, which have few number of channels and conventional multichannel BSS algorithms have failed [72, 73]. For illustration, a sample data adapted from the *abdominal and direct fetal electrocardiogram* (ADFECG)

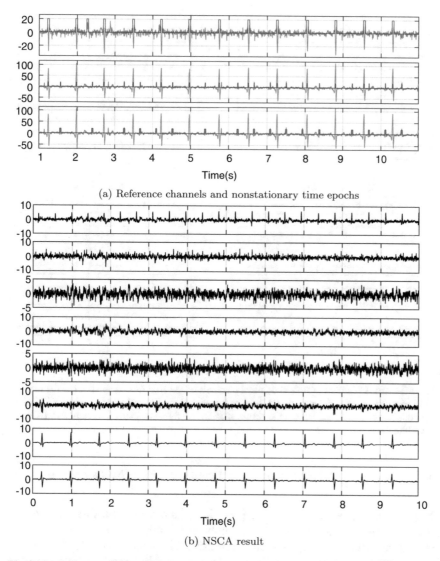

(a) Reference channels and nonstationary time epochs

(b) NSCA result

Fig. 5.23 The result of NSCA for the sample data of Fig. 5.18. (**a**) The reference mECG local power envelope time epochs (top panel), an abdominal channel local power envelope time epochs (middle panel), and the merged local power envelope time epochs after excluding the mECG time epochs (bottom panel). The nonstationary epochs are shown as red pulses. (**b**) The NSCA result. (Adapted from [42])

database [72] is shown in Fig. 5.24. As shown in this figure, the maternal and fetal ECG were not fully separable by applying JADE on the original four channels, since traces of the mECG exist in the fECG component. However, by adding an auxiliary channel according to the procedure detailed in [41], JADE has achieved in fully separating the mECG and fECG.

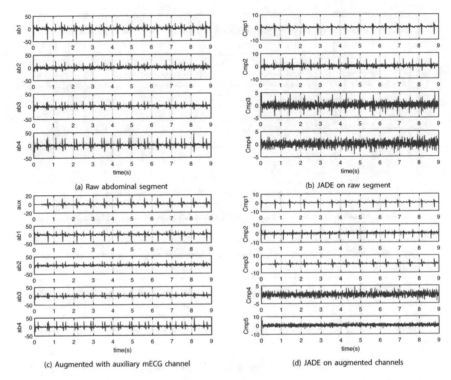

Fig. 5.24 (**a**) A segment of four abdominal channels of the ADFECG database; (**b**) the result of JADE on the original data segment; (**c**) the data segment augmented with an auxiliary mECG channel added as the first channel; (**d**) the result of JADE on the augmented data segment. (Adapted from [40, 41])

5.6.2 Maternal-Fetal Subspace Decomposition by Deflation

In [74, 85, 86], a deflation-based procedure, known as *denoising by deflation* (DEFL), was proposed for the general problem of rank-deficient and noisy source separation, with special interest in noninvasive fECG extraction. DEFL is a subspace denoising algorithm, which separates the undesired signals of multichannel noisy data using a sequence of *linear decomposition, denoising,* and *linear recomposition* in successive iterations. The overall block diagram of DEFL for mECG cancellation is shown in Fig. 5.25. Accordingly, the linear decomposition unit is generally a GEVD procedure such as πCA (or NSCA), using the R-peaks of the mother. The outputs of this unit are ranked in descending (ascending) order of resemblance with the signal (noise) subspace. This block concentrates the components of the maternal subspace in the first few components of its output. The unit is followed by a linear or nonlinear monotonic denoising filter that is applied to the first L components ($1 \leq L < N$) of the previous block. This filter can be any of the single-channel filters detailed in Sect. 5.4, applied to each channel separately, or a

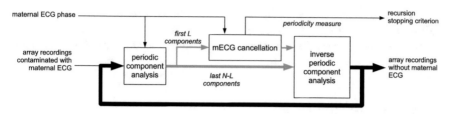

Fig. 5.25 The DEFL algorithm for separating the mECG from abdominal recordings in highly noisy and rank-deficient scenarios [85, 86]

multichannel filter applied to the first L components together. However, such denoising could have been directly applied to the original data $\mathbf{x}(t)$, but by applying it after the linear decomposition step, we benefit from the improved signal quality of the first few components extracted by the linear decomposition block. This improvement is the direct consequence of maximizing the πCA or NSCA cost functions during the GEVD procedure. Finally, the residual signals of the L denoised components and the other $N - L$ unchanged components are transformed back to the observation space, using the inverse of the linear decomposition matrix. In each iteration of the algorithm, portions of the mECG, fECG, and noise subspaces are separated, and the procedure is repeated until the output signals satisfy some predefined measure of signal/noise separability.

According to Fig. 5.25, each iteration of DEFL can be summarized as follows:

$$\mathbf{y}_i(t) = \mathbf{W}_i^{-T}\mathbf{G}\left(\mathbf{W}_i^T\mathbf{x}_i(t); L\right) \tag{5.25}$$

where $\mathbf{x}_i(t)$ is the input data of the ith iteration ($\mathbf{x}_1(t) = \mathbf{x}(t)$), $\mathbf{y}_i$ is the output of the iteration, $\mathbf{G}(\cdot; L)$ is the denoising operator applied to the first L channels of the input, and $\mathbf{W}_i$ is the spatial filter (πCA or NSCA).

The concept behind (5.25) is analogous to *wavelet shrinkage* used for single-channel denoising. An important property of the DEFL algorithm is that, unlike most ICA-based denoising schemes, the data dimensionality is preserved. Moreover, due to the denoising block between the linear projection stages, it overall performs as a nonlinear filtering scheme, which can deal with full-rank and even non-additive noise mixtures. An adaptive version of this algorithm has also been developed for online fECG extraction [33].

5.6.3   Block-Wise and Online Fetal ECG Extraction

For long multichannel data records, the batch processing requires a huge amount of memory and processing time. Moreover, during long recording sessions, it often happens that the fetus moves, which means that the fetal position changes with respect to the fixed maternal abdominal sensors. Therefore, stationary source separation algorithms, which as in (5.10) assume a constant mixing matrix **A**, will fail or result in partial fECG source separation. To resolve this issue, the data is partitioned and processed block-wise, or algorithms specific for online processing are used.

5.6.3.1 Block-Wise Analysis

Depending on the application, the maternal abdominal data can be partitioned into overlapping or non-overlapping blocks, and any of the fECG extraction schemes detailed in previous sections is applied to each block. This is the most popular method, as it does not require any change in the algorithms used for fECG extraction. However, the challenge is how to automatically identify and recombine the extracted fECG of successive blocks. Especially, as noted in Sect. 5.5, ICA algorithms, which are one of the most popular methods for fECG extraction, do not guarantee to preserve the order and amplitude of the sources over different data blocks. As a result, for non-supervised algorithms, a post-fECG extraction unit is required, which automatically detects, normalizes, and aligns the fECG of successive blocks. Automatic signal quality indexes have been proposed in the literature for adult ECG signal quality assessment [3, 22, 53]. In [41], several signal quality indexes were specifically proposed for the fECG and successfully tested over several available datasets.

5.6.3.2 Online Source Separation

An alternative solution for processing long fECG data records is to use sample-wise online source separation algorithms. Adaptive source separation algorithms are well known in the blind source separation literature. One of the most popular algorithms in this context is known as *equivariant adaptive source separation* via *independence* (EASI) [16]. In this method, the separating matrix at time instant t, denoted by $\mathbf{B}(t)$, is adaptively updated using an equivariant serial update:

$$\mathbf{B}(t+1) = \mathbf{B}(t) - \lambda(t)\mathbf{H}\big(\mathbf{y}(t)\big)\mathbf{B}(t) \qquad (5.26)$$

where $\lambda(t)$ is an update factor, $\mathbf{y}(t) = \mathbf{B}(t)\mathbf{x}(t)$ is the adaptive estimate of the independent sources, and $\mathbf{H}(\cdot)$ is a nonlinear function of the estimated sources *cumulants* [16]. For time-varying mixtures, the mixing matrix **A** defined in (5.10) becomes time-variant and the algorithm seeks the demixing matrix such that $\mathbf{B}(t)\mathbf{A}(t)$ approaches identity, i.e., where $\|\mathbf{H}(\mathbf{y}(t))\| \to 0$. This approach also works for the cases, in which the variations are due to the sources rather than the mixture. For instance, suppose that the mixing matrix $\mathbf{A}(t) = \mathbf{A}$ is constant, but the sources are nonstationary. As a result, the function $\mathbf{H}(\cdot)$ deviates from zero as the recursion reaches the nonstationary epochs of the signals. Various source separation algorithms, which use sample-wise updates (instead of global averaging), can be used for online fECG extraction [38, Ch. 3.2], [26, Ch. 4.5].

Finally note that for an online implementation of GEVD-based algorithms (such as πCA and NSCA), the covariance matrices $\mathbf{C}_x$ and $\mathbf{C}_m$ are both updated in time, i.e.,

$$
\begin{aligned}
\mathbf{C}_x(t) &= \alpha\mathbf{C}_x(t-1) + \mathbf{x}(t)\mathbf{x}(t)^T \\
\mathbf{C}_m(t) &= \beta\mathbf{C}_m(t-1) + \mathbf{x}(t)\mathbf{x}(t-\tau_t)^T
\end{aligned}
\tag{5.27}
$$

where $\alpha, \beta \leq 1$ are forgetting factors and τ_t is the time-variant heart-rate period defined in (5.4). Further details regarding the various online variants of fECG extraction algorithms can be followed from [12, 33, 41].

5.7 Fetal ECG Post-processing

5.7.1 Fetal R-Peak Detection

After extracting the fECG from maternal abdominal recordings, the next step is to extract clinical parameters from the fECG. The fetal heart rate (fHR) is the first and most commonly used parameter, which in turn requires the detection of the fECG R-peaks. In this context, classical R-peak detectors, such as local peak search over sliding windows or the well-known Pan-Tompkins method used in adult ECG [68], are the most common. However, considering the relatively low SNR of the fECG and its limited morphological shapes, specific fetal R-peak detectors have been developed that are robust to noise [12, 41]. These methods are based on a *matched filter* using fixed or adaptive QRS-like templates. A wide range of these techniques were studied and compared with one another during the annual Physionet/ Computing in Cardiology Challenge 2013 [90].

After fetal R-peaks, the fHR time series is commonly post-processed to refine the calculated heart rate time series and to correct the excess and missing R-peaks. These corrections have been commonly performed by rule-based methods, which correct the outlier R-peaks (and the corresponding heart beats), while keeping the normal beats unchanged [30, 40, 90].

5.7.2 Fetal ECG Enhancement

Depending on the signal quality, after mECG cancellation, the fECG might be directly detectable from one or more of the residual channels, or additional stages may be required for extracting the fECG from the residual background noise. As detailed in Sect. 5.4, numerous techniques have been proposed for ECG denoising, including Kalman filters [78, 79, 82], wavelet denoisers [44, 75], filtering using piecewise smoothness priors [75], etc. An example of such post-processing for fECG enhancement is demonstrated in Fig. 5.14.

For morphological analysis due to the relatively low SNR of fECG signals— even after mECG and background noise cancellation— the SNR improvement obtained by post-processing filters can still be insufficient for reliable fECG parameter extraction. In this case, an effective approach is to use synchronous weighted averaging of successive beats [52]. This procedure is known to improve the SNR by a factor of K, where K is the number of averaged beats.

5.7.3 Fetal ECG Morphological Parameter Extraction

To date, the morphological parameters of the fECG and their relationship with the well-being of the fetus are still under study. Researchers have extracted parameters such as the QT-interval [7, 9, 23] and the ST-segment [21, 60]. The typical benchmark for these studies is commonly the invasive fECG obtained from the fetal scalp electrodes acquired during labor. However, it is currently difficult to evaluate the fECG parameters independently since there are very few open-access fECG databases with expert annotations. Considering that the technology of fECG acquisition and processing is emerging as a standard procedure, it is foreseen that fetal ECG-based parameter extraction will be the main focus of research in future studies.

5.8 Conclusion

In this chapter, some of the major technologies and algorithms used for the acquisition and noninvasive processing of fetal electrocardiogram signals from maternal abdominal recordings were reviewed. The recent advances in this domain, especially during the past decade, demonstrate that the technology is emerging as a stable and reliable alternative for invasive methods. A promising future trend is to combine this technology with other low-cost fetal cardiac monitoring modalities such as the phonocardiogram (PCG) and the Doppler technology. The extension of these technologies to multiple pregnancies and pathological cases and its combination with other vital aspects such as the development of the fetal central nervous system (CNS) and cerebral growth are among the future challenges of this domain. The availability of open-access data with clinical annotations and open-source devices and algorithms are among the requirements that can significantly accelerate the development of this technology.

References

1. Andreotti, F., Riedl, M., Himmelsbach, T., Wedekind, D., Wessel, N., Stepan, H., Schmieder, C., Jank, A., Malberg, H., Zaunseder, S.: Robust fetal ECG extraction and detection from abdominal leads. Physiol. Meas. **35**(8), 1551 (2014)
2. Andreotti, F., Behar, J., Zaunseder, S., Oster, J., Clifford, G.D.: Fetal ECG Synthetic Database (FECGSYNDB) (2016). https://doi.org/10.13026/C21P4T. https://physionet.org/content/fecgsyndb/
3. Andreotti, F., Gräßer, F., Malberg, H., Zaunseder, S.: Non-invasive fetal ECG signal quality assessment for multichannel heart rate estimation. IEEE Trans. Biomed. Eng. **64**(12), 2793–2802 (2017). https://doi.org/10.1109/tbme.2017.2675543
4. Assaleh, K.: Extraction of fetal electrocardiogram using adaptive neuro-fuzzy inference systems. IEEE Trans. Biomed. Eng. **54**(1), 59–68 (2006)
5. Åström, M., Santos, E.C., Sörnmo, L., Laguna, P., Wohlfart, B.: Vectorcardiographic loop alignment and the measurement of morphologic beat-to-beat variability in noisy signals. IEEE Trans. Biomed. Eng. **47**(4), 497–506 (2000)
6. Bach, F., Jordan, M.: Beyond independent components: trees and clusters. J. Mach. Learn. Res. **4**, 1205–1233 (2003). http://cmm.ensmp.fr/bach/bach03a.pdf
7. Behar, J.: Extraction of clinical information from the non-invasive fetal electrocardiogram. Ph.D. thesis, Oxford University, UK (2014)
8. Behar, J., Johnson, A., Clifford, G.D., Oster, J.: A comparison of single channel fetal ECG extraction methods. Ann. Biomed. Eng. **42**(6), 1340–1353 (2014)
9. Behar, J., Wolfberg, A., Zhu, T., Oster, J., Niksch, A., Mah, D., Chun, T., Greenberg, J., Tanner, C., Harrop, J., et al.: Evaluation of the fetal QT interval using non-invasive fetal ECG technology. Am. J. Obstet. Gynecol. **210**(1), S283–S284 (2014)
10. Belouchrani, A., Abed-Meraim, K., Cardoso, J.F., Moulines, E.: A blind source separation technique using second-order statistics. IEEE Trans. Signal Process. **45**, 434–444 (1997)
11. Ben-Arie, J., Rao, K.: Nonorthogonal representation of signals by Gaussians and Gabor functions. IEEE Trans. Circuits Syst. II, Analog Digit. Signal Process. **42**(6), 402–413 (1995)
12. Biglari, H., Sameni, R.: Fetal motion estimation from noninvasive cardiac signal recordings. Physiol. Meas. **37**(11), 2003–2023 (2016). http://stacks.iop.org/0967-3334/37/i=11/a=2003
13. Cardoso, J.F.: Multidimensional independent component analysis. In: Proceedings of the IEEE International Conference on Acoustics, Speech, and Signal Processing (ICASSP'98), vol. 4, pp. 1941–1944 (1998)
14. Cardoso, J.F.: High-order contrasts for independent component analysis. Neural Comput. **11**(1), 157–192 (1999). https://doi.org/10.1162/089976699300016863
15. Cardoso, J.F.: Source codes for blind source separation and independent component analysis. http://www2.iap.fr/users/cardoso/
16. Cardoso, J.F., Laheld, B.: Equivariant adaptive source separation. Signal Process., IEEE Trans. **44**(12), 3017–3030 (1996). https://doi.org/10.1109/78.553476
17. Cardoso, J.F., Souloumiac, A.: Blind beamforming for non Gaussian signals. IEE – Proc. -F. **140**, 362–370 (1993)
18. Clifford, G.D.: A novel framework for signal representation and source separation. J. Biol. Syst. **14**(2), 169–183 (2006)
19. Clifford, G.D., McSharry, P.E.: A realistic coupled nonlinear artificial ECG, BP, and respiratory signal generator for assessing noise performance of biomedical signal processing algorithms. Proc. SPIE Int. Symp. Fluctuations Noise. **5467**(34), 290–301 (2004)
20. Clifford, G.D., Shoeb, A., McSharry, P.E., Janz, B.A.: Model-based filtering, compression and classification of the ECG. Int. J. Bioelectromagnetism. **7**(1), 158–161 (2005)
21. Clifford, G.D., Sameni, R., Ward, J., Robertson, J., Pettigrew, C., Wolfberg, A.: Comparing the fetal ST-segment acquired using a FSE and abdominal sensors. Am. J. Obstet. Gynecol. **201**(6), S242 (2009)

22. Clifford, G.D., Lopez, D., Li, Q., Rezek, I.: Signal quality indices and data fusion for determining acceptability of electrocardiograms collected in noisy ambulatory environments. In: Computing in Cardiology, 2011, pp. 285–288. IEEE (2011)
23. Clifford, G.D., Sameni, R., Ward, J., Robinson, J., Wolfberg, A.J.: Clinically accurate fetal ECG parameters acquired from maternal abdominal sensors. Am. J. Obstet. Gynecol. **205**(1), 47.e1–47.e5 (2011)
24. Commission, I.E.: Medical electrical equipment - Part 2-25: Particular requirements for the basic safety and essential performance of electrocardiographs. Standard, International Standard (2011)
25. Comon, P.: Supervised classification, a probabilistic approach. In: Verleysen, M. (ed.) ESANN-European Symposium on Artificial Neural Networks, pp. 111–128. D facto Publication, Brussels (1995). [invited paper]
26. Comon, P., Jutten, C. (eds.): Handbook of Blind Source Separation: Independent Component Analysis and Applications. Elsevier Science, Amsterdam (2010)
27. Cremer, M.: Über die Direkte Ableitung der Aktionstrome des Menschlichen Herzens vom Oesophagus und Über das Elektrokardiogramm des Fetus. Münchener Medizinische Wochenschrift. **53**, 811–813 (1906)
28. De Lathauwer, L., De Moor, B., Vandewalle, J.: Fetal electrocardiogram extraction by blind source subspace separation. IEEE Trans. Biomed. Eng. **47**, 567–572 (2000). https://doi.org/10.1109/10.841326
29. De Moor, B.: Database for the Identification of Systems (DaISy). http://homes.esat.kuleuven.be/ smc/daisy/
30. Dessì, A., Pani, D., Raffo, L.: An advanced algorithm for fetal heart rate estimation from non-invasive low electrode density recordings. Physiol. Meas. **35**(8), 1621 (2014)
31. Devedeux, D., Marque, C., Mansour, S., Germain, G., Duchêne, J.: Uterine electromyography: a critical review. Am. J. Obstet. Gynecol. **169**(6), 1636–1653 (1993)
32. Farvet, A.G.: Computer matched filter location of fetal R-waves. Med. Biol. Eng. **6**(5), 467–475 (1968)
33. Fatemi, M., Sameni, R.: An online subspace denoising algorithm for maternal ECG removal from fetal ECG signals. Iranian J. Sci. Technol., Trans. Electr. Eng. **41**(1), 65–79 (2017)
34. Ferreol, A., Chevalier, P.: On the behavior of current second and higher order blind source separation methods for cyclostationary sources. IEEE Trans. Signal Process. **48**, 1712–1725 (2000)
35. Gardner, W.A.: Cyclostationarity in communications and signal processing. Tech. Rep., DTIC Document (1994)
36. Hall, J., Hall, J., Guyton, A.: Guyton & Hall Physiology Review Guyton Physiology Series. Elsevier Saunders (2006)
37. Hamerling, S., Meinecke, F., Müller, K.R.: Analysing ICA components by injecting noise. In: Proceedings of the 4th International Symposium on Independent Component Analysis and Blind Source Separation (ICA2003), pp. 149–154, Nara (2003). http://www.lis.inpg.fr/pages_perso/bliss/deliverables/d19.html
38. Hyvarinen, A., Karhunen, J., Oja, E.: Independent Component Analysis. Wiley-Interscience (2001)
39. Jafarnia-Dabanloo, N., McLernon, D., Zhang, H., Ayatollahi, A., Johari-Majd, V.: A modified Zeeman model for producing HRV signals and its application to ECG signal generation. J. Theor. Biol. **244**(2), 180–189 (2007)
40. Jamshidian-Tehrani, F.: Online noninvasive fetal cardiac signal extraction. Ph.D. thesis, Artificial Intelligence, School of Electrical & Computer Engineering, Shiraz University (2019)
41. Jamshidian-Tehrani, F., Sameni, R.: Fetal ECG extraction from time-varying and low-rank noninvasive maternal abdominal recordings. Physiol. Meas. **39**(12), 125008 (2018)
42. Jamshidian-Tehrani, F., Sameni, R., Jutten, C.: Temporally nonstationary component analysis; application to noninvasive fetal electrocardiogram extraction. IEEE Trans. Biomed. Eng. (2019). https://doi.org/10.1109/TBME.2019.2936943

43. John, W., Clark, J.: Chap. 4: The origin of biopotentials. In: Webster, J.G. (ed.) Medical Instrumentation: Application and Design. Wiley, Hoboken (2009)
44. Kabir, M.A., Shahnaz, C.: Denoising of ECG signals based on noise reduction algorithms in EMD and wavelet domains. Biomed. Signal Process. Control. **7**(5), 481–489 (2012)
45. Kanjilal, P., Kanjilal, P., Palit, S., Saha, G.: Fetal ECG extraction from single-channel maternal ECG using singular value decomposition. IEEE Trans. Biomed. Eng. **44**(1), 51–59 (1997). https://doi.org/10.1109/10.553712
46. Kester, W., Engineeri, A.D.I.: Data Conversion Handbook. Newnes, Burlington (2005)
47. Khamene, A., Negahdaripour, S.: A new method for the extraction of fetal ECG from the composite abdominal signal. IEEE Trans. Biomed. Eng. **47**(4), 507–516 (2000). https://doi.org/10.1109/10.828150
48. Laguna, P., Jané, R., Olmos, S., Thakor, N.V., Rix, H., Caminal, P.: Adaptive estimation of QRS complex wave features of ECG signal by the Hermite model. Med. Biol. Eng. Comput. **34**(1), 58–68 (1996)
49. Laheld, B., Cardoso, J.F.: Adaptive source separation without pre-whitening. In: EUSIPCO-94 - The 7th European Signal Processing Conference, pp. 183–186, Edinburgh (1994)
50. Larks, S.D.: Present status of fetal electrocardiography. Bio-Med. Electron, IRE Trans. **9**(3), 176–180 (1962). https://doi.org/10.1109/TBMEL.1962.4322994
51. Leski, J.: Robust weighted averaging [of biomedical signals]. IEEE Trans. Biomed. Eng. **49**(8), 796–804 (2002). https://doi.org/10.1109/TBME.2002.800757
52. Leski, J., Gacek, A.: Computationally effective algorithm for robust weighted averaging. IEEE Trans. Biomed. Eng. **51**(7), 1280–1284 (2004). https://doi.org/10.1109/TBME.2004.827953
53. Li, Q., Mark, R.G., Clifford, G.D.: Robust heart rate estimation from multiple asynchronous noisy sources using signal quality indices and a Kalman filter. Physiol. Meas. **29**(1), 15 (2007)
54. Li, R., Frasch, M.G., Wu, H.T.: Efficient fetal-maternal ECG signal separation from two channel maternal abdominal ECG via diffusion-based channel selection. Front. Physiol. **8**, 277 (2017)
55. Liu, G., Luan, Y.: An adaptive integrated algorithm for noninvasive fetal ECG separation and noise reduction based on ICA-EEMD-WS. Med. Biol. Eng. Comput. **53**(11), 1113–1127 (2015). https://doi.org/10.1007/s11517-015-1389-1
56. Ma, Y., Xiao, Y., Wei, G., Sun, J.: A multichannel nonlinear adaptive noise canceller based on generalized FLANN for fetal ECG extraction. Meas. Sci. Technol. **27**(1), 015703 (2015). https://doi.org/10.1088/0957-0233/27/1/015703
57. Malmivuo, J.: Chap. 16: Biomagnetism. In: Bronzino, J.D. (ed.) The Biomedical Engineering Handbook, 2nd edn. CRC Press LLC, Boca Raton (2000)
58. Malmivuo, J.A., Plonsey, R.: Bioelectromagnetism, Principles and Applications of Bioelectric and Biomagnetic Fields. Oxford University Press (1995). http://butler.cc.tut.fi/malmivuo/bem/bembook
59. Martens, S.M., Rabotti, C., Mischi, M., Sluijter, R.J.: A robust fetal ECG detection method for abdominal recordings. Physiol. Meas. **28**(4), 373 (2007). https://doi.org/10.1088/0967-3334/28/4/004
60. McDonnell, C., Clifford, G., Sameni, R., Ward, J., Robertson, J., Wolfberg, A.: Comparison of abdominal sensors to a fetal scalp electrode for fetal ST analysis during labor. Am. J. Obstet. Gynecol. **204**(1), S256 (2011)
61. McSharry, P.E., Clifford, G.D., Tarassenko, L., Smith, L.A.: A dynamic model for generating synthetic electrocardiogram signals. IEEE Trans. Biomed. Eng. **50**, 289–294 (2003)
62. Meinecke, F., Ziehe, A., Kawanabe, M., Müller, K.R.: A resampling approach to estimate the stability of one-dimensional or multidimensional independent components. IEEE Trans. Biomed. Eng. **49**(12) Pt 2, 1514–1525 (2002)
63. Narimani, H.: Application of Kalman and H-∞ filters in electrocardiogram denoising. Master's thesis, Biomedical Engineering, School of Electrical & Computer Engineering, Shiraz University (2014)

64. Narimani, H., Sameni, R.: Electrocardiogram denoising using H-infinity filters. In: Electrical Engineering (ICEE), 2015 23rd Iranian Conference on (2015)
65. Niknazar, M., Rivet, B., Jutten, C.: Fetal ECG extraction by extended state Kalman filtering based on single-channel recordings. IEEE Trans. Biomed. Eng. **60**(5), 1345–1352 (2013)
66. Oostendorp, T.: Modeling the fetal ECG. Ph.D. dissertation, K. U. Nijmegen, The Netherlands (1989). http://hdl.handle.net/2066/113606
67. Outram, N.J., Ifeachor, E.C., Eetvelt, P.W.J.V., Curnow, J.S.H.: Techniques for optimal enhancement and feature extraction of fetal electrocardiogram. IEE Proc.-Sci. Meas. Technol. **142**, 482–489 (1995)
68. Pan, J., Tompkins, W.J.: A real-time QRS detection algorithm. IEEE Trans. Biomed. Eng. **BME-32**(3), 230–236 (1985). https://doi.org/10.1109/TBME.1985.325532
69. Park, Y., Lee, K., Youn, D., Kim, N., Kim, W., Park, S.: On detecting the presence of fetal R-wave using the moving averaged magnitude difference algorithm. IEEE Trans. Biomed. Eng. **39**(8), 868–871 (1992). https://doi.org/10.1109/10.148396
70. Parra, L., Sajda, P.: Blind source separation via generalized eigenvalue decomposition. J. Mach. Learn. Res. **4**, 1261–1269 (2003)
71. Pham, D.T., Cardoso, J.F.: Blind separation of instantaneous mixtures of nonstationary sources. IEEE Trans. Signal Process. **49**(9), 1837–1848 (2001)
72. PhysioNet: Abdominal and direct fetal electrocardiogram database. National Institutes of Health. https://physionet.org/physiobank/database/adfecgdb/
73. PhysioNet: Noninvasive fetal ECG database. National Institutes of Health. physionet.org/pn3/nifecgdb/
74. Sameni, R.: Extraction of fetal cardiac signals from an array of maternal abdominal recordings. Ph.D. thesis, Sharif University of Technology – Institut National Polytechnique de Grenoble (2008). http://www.sameni.info/Publications/Thesis/PhDThesis.pdf
75. Sameni, R.: Online filtering using piecewise smoothness priors: application to normal and abnormal electrocardiogram denoising. Signal Process. **133**(4), 52–63 (2017). https://doi.org/10.1016/j.sigpro.2016.10.019
76. Sameni, R.: The Open-Source Electrophysiological Toolbox (OSET), version 3.14 (2018). https://gitlab.com/rsameni/OSET/
77. Sameni, R., Clifford, G.D.: A review of fetal ECG signal processing; issues and promising directions. Open Pacing Electrophysiol Ther J (TOPETJ). **3**, 4–20 (2010). https://doi.org/10.2174/1876536X01003010004
78. Sameni, R., Shamsollahi, M.B., Jutten, C.: Filtering electrocardiogram signals using the extended Kalman filter. In: Proceedings of the 27th Annual International Conference of the IEEE Engineering in Medicine and Biology Society (EMBS), pp. 5639–5642, Shanghai (2005). https://doi.org/10.1109/IEMBS.2005.1615765
79. Sameni, R., Shamsollahi, M.B., Jutten, C., Babaie-Zadeh, M.: Filtering noisy ECG signals using the extended Kalman filter based on a modified dynamic ECG model. In: Proceedings of the 32nd Annual International Conference on Computers in Cardiology, pp. 1017–1020, Lyon (2005)
80. Sameni, R., Jutten, C., Shamsollahi, M.B.: What ICA provides for ECG processing: application to noninvasive fetal ECG extraction. In: Proceedings of the International Symposium on Signal Processing and Information Technology (ISSPIT'06), pp. 656–661. Vancouver, Canada (2006)
81. Sameni, R., Clifford, G.D., Jutten, C., Shamsollahi, M.B.: Multichannel ECG and noise modeling: application to maternal and fetal ECG signals. EURASIP J. Adv. Signal Process. **2007**, 43407, 14 pages (2007). https://doi.org/10.1155/2007/43407
82. Sameni, R., Shamsollahi, M.B., Jutten, C., Clifford, G.D.: A nonlinear Bayesian filtering framework for ECG denoising. IEEE Trans. Biomed. Eng. **54**(12), 2172–2185 (2007). https://doi.org/10.1109/TBME.2007.897817
83. Sameni, R., Jutten, C., Shamsollahi, M.B.: Multichannel electrocardiogram decomposition using periodic component analysis. IEEE Trans. Biomed. Eng. **55**(8), 1935–1940 (2008). https://doi.org/10.1109/TBME.2008.919714

84. Sameni, R., Shamsollahi, M.B., Jutten, C.: Model-based Bayesian filtering of cardiac contaminants from biomedical recordings. Physiol. Meas. **29**(5), 595–613 (2008). https://doi.org/10.1088/0967-3334/29/5/006

85. Sameni, R., Jutten, C., Shamsollahi, M., Clifford, G.: Extraction of Fetal Cardiac Signals (2010). US Patent

86. Sameni, R., Jutten, C., Shamsollahi, M.B.: A deflation procedure for subspace decomposition. IEEE Trans. Signal Process. **58**(4), 2363–2374 (2010)

87. Saul, L.K., Allen, J.B.: Periodic component analysis: an Eigenvalue method for representing periodic structure in speech. In: NIPS, pp. 807–813 (2000). http://www.cs.cmu.edu/Groups/NIPS/00papers-pub-on-web/SaulAllen.pdf

88. Scher, A.M., Young, A., Meredith, W.M.: Factor analysis of the electrocardiogram: test of electrocardiographic theory: normal hearts. Circ. Res. **8**(3), 519–526 (1960)

89. Shao, M., Barner, K., Goodman, M.: An interference cancellation algorithm for noninvasive extraction of transabdominal fetal electroencephalogram (TaFEEG). IEEE Trans. Biomed. Eng. **51**(3), 471–483 (2004). https://doi.org/10.1109/TBME.2003.821011

90. Silva, I., Behar, J., Sameni, R., Zhu, T., Oster, J., Clifford, G.D., Moody, G.B.: Noninvasive fetal ECG: the physionet/computing in cardiology challenge 2013. In: Computing in Cardiology Conference (CinC), 2013, pp. 149–152. IEEE (2013)

91. Snowden, S., Simpson, N.A., Walker, J.J.: A digital system for recording the electrical activity of the uterus. Physiol. Meas. **22**(4), 673–679 (2001)

92. Sörnmo, L.: Vectorcardiographic loop alignment and morphologic beat-to-beat variability. IEEE Trans. Biomed. Eng. **45**(12), 1401–1413 (1998)

93. Sörnmo, L., Borjesson, P.O., Nygards, M.E., Pahlm, O.: A method for evaluation of QRS shape features using a mathematical model for the ECG. IEEE Trans. Biomed. Eng. **BME-28**(10), 713–717 (1981)

94. Stinstra, J.: Reliability of the fetal magnetocardiogram. Ph.D. thesis, University of Twente, Enschede, The Netherlands (2001). http://doc.utwente.nl/35964/

95. Stogbauer, H., Kraskov, A., Astakhov, S.A., Grassberger, P.: Least dependent component analysis based on mutual information. Phys. Rev. E. **70**, 066123 (2004). https://doi.org/10.1103/PhysRevE.70.066123

96. Strang, G.: Linear algebra and its applications, 3rd edn. Brooks/Cole, Thomson Learning, USA (1988).

97. Swarnalatha, R., Prasad, D.: A novel technique for extraction of FECG using multi stage adaptive filtering. J. Appl. Sci. **10**(4), 319–324 (2010)

98. Texas Instruments: Analog front-end design for ECG systems using delta-sigma ADCs (2010)

99. Tong, L., Liu, R.W., Soon, V., Huang, Y.F.: Indeterminacy and identifiability of blind identification. IEEE Trans. Circuits Syst. **38**, 499–509 (1991)

100. van Trees, H.: Detection, Estimation, and Modulation Theory. Part I. Wiley, New York (2001)

101. Vigneron, V., Paraschiv-Ionescu, A., Azancot, A., Sibony, O., Jutten, C.: Fetal electrocardiogram extraction based on non-stationary ICA and wavelet denoising. In: Signal Processing and Its Applications, 2003. Proceedings. Seventh International Symposium on **2**, pp. 69–72, vol. 2 (1–4 July 2003). https://doi.org/10.1109/ISSPA.2003.1224817

102. Webster, J.G. (ed.): Medical Instrumentation: Application and Design, 3rd edn. Wiley, New York (1998)

103. Weiss, Y.: Segmentation using eigenvectors: a unifying view. In: Proceedings IEEE International Conference on Computer Vision (2), pp. 975–982 (1999). citeseer.ist.psu.edu/weiss99segmentation.html

104. Widrow, B., Glover, J., McCool, J., Kaunitz, J., Williams, C., Hearn, H., Zeidler, J., Dong, E., Goodlin, R.: Adaptive noise cancelling: principles and applications. Proc. IEEE. **63**(12), 1692–1716 (1975)

105. Yeredor, A.: On optimal selection of correlation matrices for matrix-pencil-based separation. In: International Conference on Independent Component Analysis and Signal Separation, pp. 187–194. Springer, Berlin, Heidelberg (2009)

106. Yeredor, A.: Chapter 7: Second-order methods based on color. In: Handbook of Blind Source Separation, pp. 227–279. Elsevier (2010). https://www.elsevier.com/books/handbook-of-blind-source-separation/comon/978-0-12-374726-6
107. Yeredor, A.: Performance analysis of GEVD-based source separation with second-order statistics. IEEE Trans. Signal Process. **59**(10), 5077–5082 (2011)
108. Zarzoso, V., Nandi, A.: Noninvasive fetal electrocardiogram extraction: blind separation versus adaptive noise cancellation. IEEE Trans. Biomed. Eng. **48**(1), 12–18 (2001). https://doi.org/10.1109/10.900244

Chapter 6
Innovative Devices and Techniques for Multimodal Fetal Health Monitoring

M. J. Rooijakkers

Contents

6.1 Introduction

Over the last couple of decades, a clear trend in clinical care is visible, in which care is moved from the hospital and healthcare centers to an at-home setting [1]. This is in part driven by the need to reduce healthcare costs, where new ambulatory monitoring techniques, combined with telehealth, could help hospitals and healthcare providers bend the cost curve [2]. In addition, locating services in a patient's home makes care easily accessible and it leads to more comprehensive and effective care, moving the emphasis towards early diagnosis and healthcare rather than disease care. Or, in the words of Mark Bertolini, CEO of the US-based insurance company Aetna: "If you have to go to the hospital, we have failed you."

In recent years, this trend of moving toward at-home care has accelerated with the availability of new and better healthcare technologies suitable for at-home use. Personalized healthcare coaching smart apps such as Sanitas' HealthCoach are

M. J. Rooijakkers (✉)
Bloomlife, Genk, Belgium
e-mail: michiel@bloomlife.com

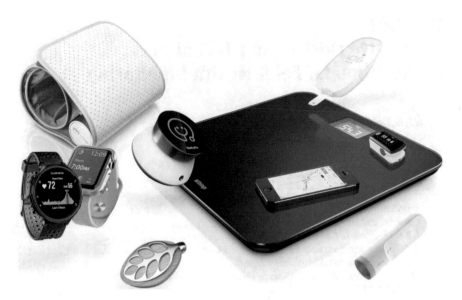

Fig. 6.1 An overview of commonly used smart sensors, including smart-watches with heart-rate and activity tracking, an activity tracking jewel, a blood pressure cuff, stethoscope, scales, glucose meter, pulse-oximeter, and body thermometer, all of which allow for at-home health monitoring

becoming more popular, and medical diagnostics apps such as Insight Optics and FibriCheck are already gaining traction. This is a trend which can only benefit from platforms such as Apple's HealthKit, the ongoing developments in artificial intelligence, and the increasing presence of personal health sensors and devices. Figure 6.1 shows an overview of personal connected devices for home use, which allow capturing one or more health-related parameters and, combined, can give insight into the user's health. The use of these types of devices is now commonplace and both accepted and utilized by the medical community to improve healthcare [3].

Home monitoring of fetal health as part of standard care, however, is currently limited to maternal perception and counting of fetal movements [4], as currently, continuous fetal monitoring devices are only available for high-risk pregnancies and require interpretation by a medical professional to give any insight into the fetal health. Instead, pregnancy monitoring is currently performed intermittently in a hospital environment, despite increasing preference of pregnant women for monitoring at home and potential benefits for the mother and healthcare system [5, 6]. As a result, the obtained fetal health information is very sparse, unless the patient is admitted to a care center for a prolonged period of time.

As an increase in high-risk pregnancies can be observed in industrialized countries, in part because of the increasing age at which women decide to get pregnant [7], various groups are working on changing the options to improve prenatal fetal monitoring and care. Universities, start-ups, and big tech invest in research and technology in preparation for a transformation of prenatal care to new technologies and toward an at-home setting [8–10], with a number of potential big improvements over current clinical practice:

1. Continuous monitoring, instead of intermittent observations at a care center.
2. Earlier detection of possible issues enabling a preventative style of medicine.
3. Lower load on the hospital and medical personnel.
4. Reduction of overall healthcare cost, while improving outcomes.

In the remainder of this chapter, we will first go over the currently available fetal health monitoring techniques and see a shift toward new measurement techniques. In addition, an overview is provided of the devices currently available for home monitoring of fetal health features. Next, focusing on fetal motion detection, we will have a look at current research toward fetal monitoring and how to approach the signal processing and data analysis issues when developing a fetal monitoring device.

6.2 Current Fetal Health Monitoring Techniques

Currently, detection of fetal health during pregnancy is limited to the measurement of a couple of features, which can be obtained using a number of different measurement techniques. The most common and most important features used for fetal health assessment are the fetal heart rate (FHR) and counting of fetal movements [11]. Accurate FHR recordings allow determination of fetal development, fetal maturity, and the existence of fetal distress or congenital heart disease [12]. A reduction in fetal movements, on the other hand, is associated with a wide variety of pregnancy pathologies, ranging from intrauterine growth restriction to anomalies affecting neurological systems, brain injuries, and death [13]. In addition, assessment of fetal growth and anatomy abnormalities and, in some cases, information from fetal (scalp) blood sampling can provide extra insight into the fetal well-being.

6.2.1 Devices in Clinical Fetal Health Monitoring

One of the most commonly used devices in fetal health monitoring throughout pregnancy is the Doppler ultrasound (US) monitor, which was introduced in 1958 by Edward Hon, MD [11]. In Doppler US measurements, ultrasonic waves generated by a transducer are transmitted into the body, where they experience a shift in frequency when backscattered by moving targets with a different density. The magnitude and direction of the recorded shift in frequency of the backscattered waves, known as the Doppler effect, contains information about the motion of the scatterers. This way, the Doppler effect can be used to non-invasively track the movements of the fetal heart and valves. Fast shifts of these body parts can be tracked to determine the averaged FHR over time. In a similar way, slow shifts in frequency and

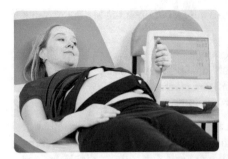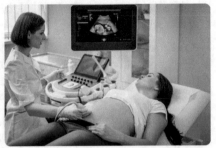

Fig. 6.2 Current CTG machine with Doppler US probe for FHR monitoring (left) and ultrasound imaging (right)

direction can be retraced to fetal body movements. Figure 6.2 shows a Doppler US system in use in the clinic. A more recent M-mode Doppler US allows measurement of tissue deformation and fluid flow and, as a result, the assessment of types of arrhythmia based on the contractile and flow behavior in the atria and ventricles.

The accuracy of the average FHR detection by Doppler US is good, although even recent Doppler US techniques are not able to provide the FHR signal accuracy required for reliable quantitative evaluation of FHR variability or detection of the presence of arrhythmias [14]. Detection of fetal motion using Doppler US is limited to strong limb and full body movements, which can be detected with an accuracy of around 94% [15, 16]. The above results assume the probe to be accurately aimed at the fetal heart, which requires the occasional attention of trained medical personnel to adjust the probe position and orientation. In addition, the Doppler US introduces energy into the body of both mother and fetus, which might pose a health risk with prolonged exposure, which makes Doppler US measurements unsuitable for long-term observation [17, 18]. Therefore, Doppler US is mostly used for intermittent checks of the FHR throughout delivery, often in conjunction with recordings of uterine activity as part of the cardiotocogram (CTG).

By using an US probe with an array of densely packed transducers, the location and movement of scatterers can be tracked over a two- or three-dimensional area, and the received backscattered signals can be converted into images or video. Imaging and video are typically made by trained medical personnel to check on the fetal growth and development. US video allows the clinician to determine the gender of the fetus, estimate its size and weight, and check for any anatomical deformations. Continuous US imaging is currently the most reliable and accurate method to monitor fetal movement when identification of the fetal movements is performed manually by a medical expert [15]. In addition, it allows for discrimination between different types of fetal movement, such as fetal breathing, hiccups, and limb and full body movements [15]. It is, therefore, considered the gold standard in fetal movement monitoring. However, both the recording and annotation processes are very labor-intensive; they require the continuous attention of a trained physician to handle the US probe and analyze the recorded video images. Because of this, and the energy introduced in the body, which is higher than that of the Doppler method, US imaging is not suitable for regular or long-term monitoring. Also, depending on the gestational age (GA) and the limited field of view, only part of the fetus can be visualized, which can result in missing or mislabeling of fetal movements.

A very accurate way to record the fetal heart rate is by means of the fetal electrocardiogram (ECG), which measures the changing electrical field generated by the fetal heart [19]. Classically, the fetal ECG is recorded, while the cervix is dilated using a measurement electrode directly connected to the fetal head, with a reference electrode on the thigh of the mother, as shown in Fig. 6.3. This results in a clean gold standard fetal ECG signal, which not only allows for FHR extraction but also enables a detailed analysis of the fetal ECG morphology. Detailed analysis of the fetal ECG can be used to detect adverse events such as prolonged oxygen deprivation using ST-segment analysis, although the reliability of this method is controversial [21]. As the measurement electrode needs to be screwed directly into the fetal scalp, this is a very invasive technique which requires the membranes to be ruptured. As such, a direct fetal ECG using a scalp electrode can only be performed during delivery and is typically only performed in case of possible fetal distress, if no trustworthy FHR signal can be obtained using alternative methods.

A method closely related to the fetal ECG is the fetal magnetocardiogram (MCG), which measures the magnetic field produced by electrical currents in the fetal heart as a result of the propagating action potentials. The MCG is a non-invasive method, which allows for detection of the FHR and analysis of the signal morphology as early as in the 21st week of gestation [22]. Although fMCG has shown to provide clinically accurate information suitable for, for example, fetal arrhythmia detection, it is not commonly used. As the magnetic fields emanating from the fetal heart are extremely small, the recording has to be performed using a bulky biomagnetometer in a magnetically shielded room, which makes the method only suitable for short-term observations in the hospital.

Fetal health monitoring is also possible using auscultation, by listening to the sounds of the fetal heart from the abdominal wall using a stethoscope, which, when recorded, is called a fetal cardiophonograph (CPG). Such a CPG can be used much in the same way as a Doppler US recording, although the signal-to-noise ratio

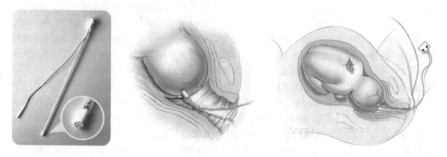

Fig. 6.3 Picture of fetal scalp electrode and graphic depiction of its placement. (Modified from [20])

(SNR) is typically worse due to the presence of extraneous sounds. It has, therefore, become a lot less popular with the advent of Doppler US [11].

A final method worth mentioning to get insight into fetal health is maternal perception, as it is not only the oldest, but also still the most commonly used method to assess fetal well-being. It is often used as a first indicator of fetal health, where a strong reduction of the amount of fetal movement felt by the mother is considered as a strong indicator for reduced fetal health [13]. Despite the availability of alternative methods to record fetal movements in the hospital, manual annotation in the hospital is also still common practice during CTG recordings, as shown on the left of Fig. 6.2.

6.2.2 New Measurement Techniques in the Clinical Environment

Recently, a shift to compacter and less invasive systems in fetal health monitoring has started. These new systems allow for longer unsupervised recordings, while providing caregivers with more information about the current fetal health state. Most noticeably, a clear trend toward the use of abdominal electrodes is visible, which allow recording of the fetal ECG and analysis of the FHR throughout pregnancy.

6.2.2.1   Fetal Heart Rate Monitoring

As mentioned in Sect. 6.2.1, the current gold standard for FHR monitoring uses the fetal ECG with a scalp electrode. An alternative way to obtain the fetal ECG is by means of electrodes placed on the maternal abdomen. This allows for recording of the fetal ECG and extraction of the FHR and morphological information noninvasively throughout all stages of pregnancy [23]. This method was used for the first time over a century ago, but the low SNR of the recorded signals has, until recently, limited its use as a diagnostic tool [17]. With improvements in electronics and especially the accurate conversion of the recorded signals to the digital domain and improvements in digital signal processors (DSP), powerful processing techniques have become available to significantly increase the SNR. Source separating and filtering techniques can now be applied in real time to multi-channel recordings to provide signals with an SNR rivaling those recorded using the fetal scalp electrode, while offering multi-lead derivations, resulting in improved interpretability of the fetal ECG [23]. It should be no surprise that recent years have seen the first couple of abdominal fetal ECG monitoring devices being approved for medical use in the clinic by the U.S. Food and Drug Administration (FDA); for example, the Monica AN24 monitor and Novii Wireless Patch System by GE Healthcare (Chicago, IL, USA), the Meridian M110 from MindChild Medical (North Andover, MA, USA),

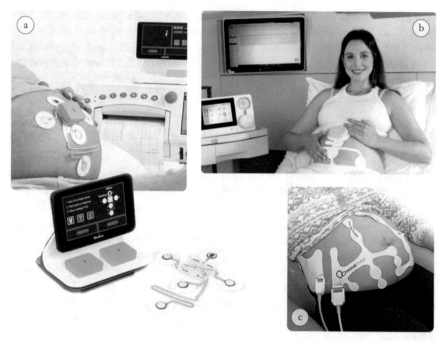

Fig. 6.4 Pictures of approved new-style medical devices for fetal health monitoring in a clinical setting, with (**a**) Novii Wireless Patch System by GE, (**b**) Nemo Fetal Monitoring System by Nemo Healthcare, and (**c**) Meridian M110 from MindChild Medical

and the Nemo Fetal Monitoring System by Nemo Healthcare (Veldhoven, NL), shown in Fig. 6.4.

The use of fetal pulse oximetry, which allows noninvasive monitoring of the oxygen saturation in the arterial blood of the fetus, is currently in the research and development stage. The pulse oximeter (POM) can be applied either on the maternal abdomen for a trans-abdominal measurement or on the fetal scalp for a direct recording. The trans-abdominal method is noninvasive but hard to calibrate and suffers from motion artifacts. The latter method is partially invasive, as it requires secure placement of the sensor on the fetus and hence dilation and rupturing of the membranes, limiting its use to labor, although no skin perforation is required like with the fetal scalp electrode [24]. Both versions of fetal POM offer continuous insight in both FHR and blood oxygenation, which promises a reduced need for caesarean sections.

6.2.2.2   Fetal Movement Monitoring

New developments in fetal movement detection are mostly limited to the research stage, where a number of different measurement techniques are explored in an effort to overcome the various limitations of the fetal movement detection methods currently available in clinical practice. These new methods are based on measurement of abdominal skin deformation and acceleration [25], optical flow displacement [26], the fetal phonogram [27], FHR accelerations [28], and morphological changes in the fetal ECG [29, 30]. So far, however, none of this research has resulted in a product suitable for clinical use.

6.2.3 At-Home Fetal Monitoring Devices

In current clinical practice, even with the use of the new generation of monitoring devices, pregnancy monitoring is performed intermittently, as the patient has to go to a care center for observation. As a result, the obtained information is relatively sparse unless the patient is admitted to the hospital for a prolonged period of time. Some of the previously mentioned measurement techniques, which show a clear move to more compact devices for fetal health monitoring, could allow for long-term, continuous monitoring of the relevant features indicative of fetal well-being in an ambulatory at-home setting. As a result, the trend of moving toward home monitoring is also taking hold in the field of fetal health monitoring, with numerous products appearing for at-home fetal monitoring. Figure 6.5 shows an overview of the various devices.

Evidently, a large number of at-home fetal monitoring devices exists and, from Fig. 6.5, it is also clear that most devices target recording of the FHR. On closer examination, we can see that the majority of the devices require either continuous manual operation or regular adjustments to function correctly. In addition, a number of these devices use an active sensing method, which introduces energy in the maternal and fetal tissues. This means that, although these devices enable recording fetal health features on a regular basis, not all are suitable for continuous long-term monitoring.

6.3 Toward Clinically Relevant At-home Fetal Health Monitoring: Fetal Motion Detection

In this chapter, we will have a closer look at the methods used and the trade-offs made in development of a fetal health monitoring device suitable for continuous unobtrusive monitoring. Because the amount of research into and the number of available devices for unobtrusive at-home FHR monitoring is much greater than those for fetal movement monitoring, while the latter is currently the most commonly used feature for at-home fetal health estimation, this chapter will focus on fetal motion detection.

Fig. 6.5 Overview of both consumer and clinical grade devices suitable for at-home monitoring of fetal health related features. From the left top, in clockwise direction: (**a**) ECG patch-based Bloomlife Beatle prototype* (bloomlife.com), two ECG wearable belt systems (**b**) Nuvo Invu* (nuvocares.com) and (**c**) Owlet pregnancy band* (owletcare.com), (**d**) Babywatcher ultra-sound video (mybabywatcher.com), (**e**) a consumer Doppler US, (**f**) HeraMED HeraBEAT™** and (**g**) EchoBEAT™* wireless Doppler US (hera-med.com), (**h**) Sense4baby portable CTG monitor (sense4baby.nl), (**i**) the Modoo* (modoo-med.com) and (**j**) XinKaishi** (mykaishi.com) Bluetooth-connected fetal stethoscopes, (**k**) a digital stethoscope for fetal CPG recordings, and (**l**) eMotion accelerometer-based fetal movement detector*

*These devices are, at the time of writing, still under development
**These devices have, at the time of writing, a limited geographical availability

6.3.1 Selection of Measurement Modalities

Before one or more measurement methods can be selected for detection of fetal motion, it is important to weigh the pros and cons of all possible measurement methods. To this end, Table 6.1 shows an overview of all methods for detection of fetal motion found in literature. Since the monitoring method should be suitable for long-term at-home use, some methods can be eliminated immediately based on one or more of their characteristics. Maternal sense, which is the current de facto standard, requires constant attention by the mother. The biomagnetometer is a very large and expensive device requiring a magnetically shielded environment. Ultrasound video requires constant attention of a trained professional and long-term use is not advised due to the energy being transmitted into the body [32]. Both Doppler ultrasound- and the fetal phonogram-based methods require regular adjustments to keep the sensor aimed at the fetal heart. The remaining modalities, based on a skin stretch sensor, accelerometer, or abdominal ECG, are all small, passive, and can be used long-term without the need for user attention.

Skin-stretch and accelerometer sensors both try to capture the same signal, which is the motion of the fetus mechanically transmitted through the maternal abdominal tissues to the mother's abdominal wall. The abdominal ECG, on the other hand, measures the electrical activity of the fetal heart and fetal movement might be detected by looking at morphological changes of the measured ECG waveforms [29]. Choosing which sensor to use will also depend on the types of fetal motion it is able to detect. Table 6.2 gives a classification of the various types of motion observed in a fetus. We have opted here to pursue the use of accelerometers to try and capture limb movements or kicks and hiccups, as these should result in discrete short abdominal movements, and the abdominal ECG to detect rolling and stretching movements by looking at the effect of the slow fetal rotational and translational movement with respect to the electrodes on the ECG morphology.

Table 6.1 Different measurement methods capable of fetal movement detection including their main characteristics [24, 31]

Measurement method	Invasiveness	Size	Detectability	Attention	Max duration
Maternal sense	None	–	High	Continuous	–
Biomagnetometer	Low	Very large	High	None	Hour
Ultrasound video	Low	Average	Very high	Continuous	Hours
Doppler ultrasound	Low	Average	Medium	Regular	Day
Fetal phonogram	None	Average	Low	Regular	–
Skin stretch sensor	None	Small	Medium	None	–
Accelerometer	None	Very small	Medium	None	–
Abdominal ECG	None	Small	Medium	None	–

Table 6.2 Different types of fetal movements and their defining characteristics

Movement type	Characteristics
Rolling movement	Rolling or rotating movements, possible change in fetal position
Stretching	Big movements of the fetal thorax without rolling motion
Limb movements	Single or repeated movement of one or more fetal limbs
Hiccups	Rhythmic twitching of the fetal body as a result of involuntary diaphragm movements
Fetal breathing	In- and exhalation of amniotic fluid

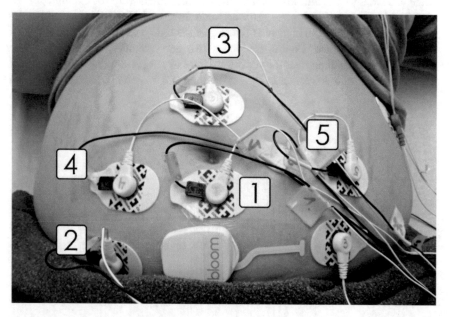

Fig. 6.6 On-body placement of the used sensors. The TMSi Porti7 sensors consist of 5 accelerometers placed on the abdomen, with a sixth accelerometer placed on the back (not visible) and 6 electrodes used to acquire abdominal ECG data. The Bloomlife sensor with built-in accelerometer and electrodes is placed directly below TMSi accelerometer 1

6.3.2   Recording a Reference Data Set

Any algorithm development, model training, or analysis of detection accuracy requires a representative dataset with accurate reference annotations. To this end, measurements were performed using two devices. Firstly, a Porti7 device from Twente Medical Systems International (TMSi) was used. The TMSi Porti7 was configured to record data from six triaxial accelerometers and six abdominal electrodes at 2048 Hz. Five of the TMSi accelerometer sensors were positioned on the abdomen with the navel serving as central marker. The sixth accelerometer was placed on the back as reference for maternal movement, as shown in Fig. 6.6. Secondly, a research version of the Bloomlife wearable was used to acquire two-channel electrophysiological (ExG) data at 4096 Hz together with data from a single triaxial accelerometer at 128 Hz. Using this setup, 114 recordings of at least 20 minutes were collected from 57 pregnant women at gestational ages from week 30 onwards. All the women were lying in a hospital bed and were given a hand-held button, which they pressed when feeling fetal movement to generate a first reference for fetal movements. Short presses were used for kicks, while the button was continuously pressed for longer full-body movements.

After some initial preprocessing to limit the frequency content of the data to a range expected to contain information on fetal movements, by for example, bandpass filtering the accelerometer data between 1 and 20 Hz with a second-order Butterworth IIR filter, and high-pass filtering of the ExG data at 0.5 Hz, we can have a first look at the data. Figure 6.7 shows two accelerometer channels and one abdominal ECG channel over the first 10 minutes of a recording, as well as manual annotations of fetal movements by the mother and maternal movements by an

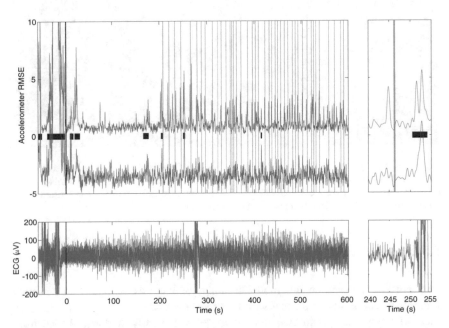

Fig. 6.7 Example accelerometer and abdominal ECG signal segments of 11 minutes (left) and 15 s (right). The top plots show RMS energy of accelerometer channels 2 and 6 with added offset in blue and red, respectively, as well as the maternal button presses (gray lines), annotated maternal motion (black box) and the official start of the measurement (black line). The bottom plots show the abdominal ECG for channel 2 after high-pass filtering at 0.5 Hz. Notice that segments with increased accelerometer energy always precede the manual button presses

external observer. We notice that all short button presses by the mother are preceded by an increase in accelerometer RMS energy in channel 2, but not in channel 6 (on the back), by approximately 1–2 s. To compensate for this varying offset, manual alignment of button presses with closely preceding RMS energy spikes was performed as an annotation post-processing step. In Fig. 6.7, we can also observe that segments marked as maternal motion typically show an increase in RMS energy in both accelerometer channels.

6.3.3 Accelerometer-Based Fetal Motion Detection

The first measurement method we will look into for detection of fetal movement is the use of accelerometers, which should allow for detection of "faster" fetal movements such as kicks. The general method of fetal movement detection will, in this case, be machine-learning based and follow the work in [25, 33]. A machine learning method was chosen because we face a classification problem where we want to determine for each moment in time whether the fetus moved or did not move. In addition, we have a large and accurately annotated dataset of representative recordings, which is a requirement for Machine learning techniques even more than for classical signal processing methods.

6.3.3.1 Base Method

Although machine learning offers a semi-automatic design methodology, several design choices need to be made, ranging from feature computation to selecting the proper performance metrics. Therefore, we will have a look at the design choices and validation techniques before we can analyze trade-offs in, for example, the number and positioning of sensors.

Features Features are calculated on 0.5 s windows, given the short duration of fetal kicks as observed in Fig. 6.7. Experimentation using longer time windows showed an averaging effect on the accelerometer signal. The selected features were: mean, standard deviation, inter-quartile range, correlation between axis of each sensor, and correlation with the reference sensor based on RMS energy. All features were computed for each sensor and each axis, for a total of 83 features.

The chosen performance metric The chosen performance metric is the F-score, which is defined as:

$$F_1 = 2 \cdot \frac{Se - PPV}{Se + PPV}, \qquad (6.1)$$

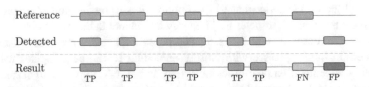

Fig. 6.8 Example of the used evaluation strategy

where Se and PPV are the sensitivity and positive predictive value, respectively. The Se and PPV are in turn defined based on the number true positive (TP), false positive (FP) and false negative (FN) detections as Se = TP "TP , FN" (fraction of kicks that is correctly identified as such) and PPV = TP "TP , FP" (fraction of detected kicks that are actually kicks). The TP, FP, and FN were calculated using the short anno-tated fetal movement events after time correction, as shown in Fig. 6.8.

The classifier The classifier was selected to be a random forest, as it offers conve-nient ways to deal with class imbalance by selecting a subset of the majority class at each iteration, similar to selecting a subset of features at each iteration. This, therefore, allows us to train our model on balanced data without discarding relevant information. After optimizing for F_1 score, we found an optimal balance for our dataset, including all data from the minority class (kicks) and one fifth of the major-ity class data.

Initial results Initial results using only a single sensor show both a Se and PPV of 0.51 for all sensors with the exception of reference sensor 6, which gave a Se and PPV of 0.0, highlighting how the latter is optimal to distinguish between maternal and fetal movements. These results are not great, as only about half of the fetal movements are captured, while the kicks which were detected have a 49% chance of having been detected erroneously.

6.3.3.2 Detection Improvements

Various techniques to improve on the initial results can be made. Some possible improvements are discussed below, including the obtained results.

Using multiple sensors Using multiple sensors at the same time might improve detections. The TMSi Porti7 recordings in the used measurement setup contain the data of six simultaneously sampled triaxial accelerometers, while so far we have only used a single accelerometer at a time. Instead of using only a single accelerom-eter, features of multiple accelerometers can be combined to detect movements more accurately. This way, the mean Se and PPV gradually increase with the num-ber of used sensors, as shown in the "No reference" columns in Table 6.3, from which we can observe a clear improvement in especially Se. The increase in Se can be explained as a reduction in missed fetal kicks, since a larger portion of the mater-

Table 6.3 The influence of the numbers of used sensors on fetal movement detection [25]

# Sensors	No reference		With reference	
	Se	PPV	Se	PPV
1	0.51	0.51	0.57	0.56
2	0.63	0.54	0.68	0.61
3	0.69	0.57	0.70	0.63
4	0.70	0.58	0.75	0.65
5	0.70	0.58	0.75	0.65

nal abdomen is covered, making sure kicks, which only result in local abdominal movements, are captured. Despite a significant increase in Se, the PPV has remained low, indicating that many maternal movements were mistaken for fetal movements.

A reference sensor A reference sensor which cannot capture any fetal movements might be used to distinguish between maternal and fetal movements. As can be seen in the "With reference" columns in Table 6.3, including features from the reference accelerometer on the back does indeed improve detection results in a more balanced way. In general, the overall improvement when adding the single reference channel is larger than including an additional abdominal sensor. The best model is now able to detect 75% of all fetal kicks, where we are 65% sure about the correctness of any detected kick.

Additional signal processing Additional signal processing might allow for a further improvement by including additional features which can capture, for example, time scale or amplitude variations. To help distinguish between the different dynamics of maternal and fetal movement, features calculated on a window of 4 s in addition to the 0.5 s window. The rationale is that short fetal movements should average out over the 4 s windows, but are captured by the 0.5 s ones, while maternal movements should appear in both. The longer window was set to 4 s, as this is long enough to average out accelerations due to fetal kicks while being short enough to limit processing delays. In addition, new features to better capture acceleration amplitude were added in the form of the sum, min, max, and magnitude of the windowed signals for all sensors and channels. Finally, all of the previously discussed features were also calculated on a single channel of the measured abdominal ExG after band-pass filtering between 0.1 and 3 Hz, which should capture any electrode motion artifacts as a result of skin stretching due to fetal kicks.

In Table 6.4, we can observe that adding the additional amplitude-based features improved fetal movement detection, especially in case of a single accelerometer. Including features on a 4 s window improved the PPV, again most noticeably in the single channel device, which does not benefit from a reference sensor to reject maternal motion. Finally, the addition of ExG features did not significantly improve the detection result.

Table 6.4 Effect of additional features, window sizes, and signal sources on fetal movement detection.

Included features	TMSi Porti7		BLoomlife	
	Se	PPV	Se	PPV
Base	0.75	0.65	0.51	0.51
+ additional features	0.76	0.71	0.65	0.65
+ 4 s window	0.75	0.76	0.64	0.75
+ ExG features	0.75	0.77	0.64	0.76

Here the TMSi Porti7 uses all six accelerometers, including one reference on the back, while the Bloomlife sensor only uses a single accelerometer [33]

6.3.3.3 Discussion

The above results show that detecting individual movements remains challenging using accelerometers, even under optimal (hospital) conditions. Especially using only a single accelerometer it might be very hard to attain a detection quality which allows for reliable detection of each individual fetal kick. From a clinical perspective it is, however, not required to count every individual kick, as long as we can gain insight into the activity level of the fetus. This can be achieved by looking at the number of detected peaks over a longer time period, e.g. 5 or 20 minutes. In Fig. 6.9, which shows the total number of detected kicks over a 20-minute period, we can see a clear correlation between the number of annotated and detected kicks for both systems. Therefore, clinically relevant insight in the amount of fetal movement can be identified with reasonable accuracy when using 5–20 minute long segments, e.g. using movement categories.

6.3.4 Fetal ECG Based Fetal Motion Detection

As we have found above, an accelerometer-based system is reasonably well-suited for detection of fetal kicks and hiccups, due to the shorter forceful type of fetal movements, but has a hard time detecting slower full-body fetal movements. Various methods have been proposed using analysis of the abdominal fetal ECG based on variations in shape and amplitude of the fetal QRS complex [29, 34, 35]. These methods build on the premise that the fetal ECG waveform as observed from the electrodes on the maternal abdomen changes as a result of a shift of the fetal cardiac vector with each movement of the fetal thorax [36]. A change in fetal QRS-wave height and shape can, therefore, be used to give an indication of the fetal motility.

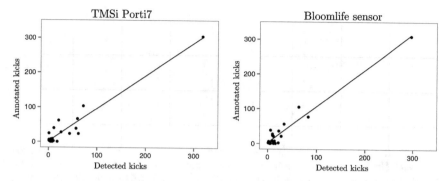

Fig. 6.9 Number of kicks detected versus number of annotated kicks for each analyzed 20-minute recording using the TMSi Porti7 (left) and Bloomlife sensor (right) using the improved models

6.3.4.1   Fetal QRS Extraction

Before analysis of the fetal QRS complex can be performed, it has to be extracted from the abdominal ExG, which in itself can be a challenging task due to the low signal amplitude and relatively high amplitude of interferers. The maternal ECG is typically the main interferer, and to remove it first all maternal R-peak locations are determined using the R-peak detection algorithm presented in [37], followed by removal of the detected maternal QRS-complexes by blanking, as shown in Fig. 6.10. Blanking reduces the number of fetal QRS-complexes whose shape might be affected by the maternal QRS enabling fetal R-peak detection. More complex methods such as those listed in [35] or [38] are possible and will increase the number of detected fetal peaks. However, these additional peaks likely contain residual errors from the maternal QRS removal which, unless removed, might introduce errors in template generation.

Fetal R-peak detection can be performed, using e.g. the method proposed in [37], on the abdominal ECG after maternal ECG removal. Cleaning of any detected fetal QRS complexes is important, as the amplitude of the fetal QRS is very low and any interference by artifacts or noise seriously impacts the fetal QRS waveform. Therefore, cleaned QRS-complexes are calculated by averaging the time-aligned fetal QRS waveforms within a 5 s interval after removing outliers which differ most in shape from the calculated average QRS complex. A 5 s window is used as a trade-of between the ability to clean and improve the QRS shape and the risk of suppressing short fetal-movement-induced changes of the QRS waveform. Figure 6.11 shows an example of the QRS complex averaging, where changes over time in the averaged fetal QRS-complex can now be used for fetal movement detection.

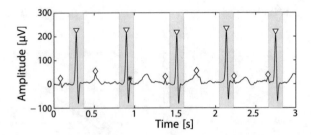

Fig. 6.10 Example of an abdominal ECG signal recorded using a single bi-polar electrode pair after basic filtering. The shaded region around each maternal R-peak (r) indicates the part of the signal which is blanked before detection of the fetal R-peaks indicated by a triangle. The diamond indicates a fetal peak which is not detected due to the blanking procedure

Fig. 6.11 Example of multiple aligned fetal QRS-complexes (thin gray) and a clean averaged fetal QRS-complex (thick black)

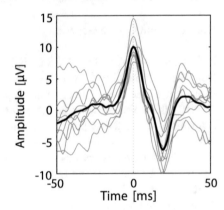

6.3.4.2 Fetal Movement Detection

Fetal movements are detected based on the premise that changes in QRS amplitude and shape are associated to translational and rotational movement of the fetus, respectively. Therefore, translational movements of the fetus with respect to the electrodes can be detected by changes in the fetal QRS amplitude over time. The translational movement feature M is calculated as the root mean squared value of all peak-to-peak fetal QRS amplitudes over a period of 10 s. A rotational movement of the fetus relative to the abdominal electrodes can be detected by calculating the correlation coefficient between consecutive QRS complexes, as it provides a measure of similarity in signal morphology. The feature for rotational movement M is, therefore, defined as the correlation coefficient between cleaned QRS complexes spaced 10 s apart [29].

Fetal movements are detected using a simple linear model and threshold which was obtained by training a linear regression model to find optimal feature weights. Different from the training and validation method for kick detection, the detected fetal movement class is compared with the reference annotations on a second-by-second basis, as the duration of fetal movements can vary greatly. The detection accuracy (Ac) defined as

$$Ac = \frac{TP + TN}{TP + FP + FN + TN},$$

$$(6.2)$$

was chosen as the performance metric, as it indicates a degree of closeness of the detections to the reference annotations. Table 6.5 shows the results of the fetal QRS shape based fetal movement detection algorithm using patient specific optimizations and features optimized for each GA age group as well as for the dataset as a whole using cross-validation.

6.3.4.3   Discussion

The patient-specific results in Table 6.5 show the optimal results we can expect using the currently used features, while the cross validation results on the whole dataset show how the algorithm would preform on a general population when using a single model. The overall Se (64%), Sp (70%), and Ac (68%) for patient optimized classification are reasonable, especially for recordings in the last weeks of gestation with an Se of 78% [39]. When training on the whole dataset, the results give an indication of the expected performance when applying the method on a new patient without training, in which case a clear drop in detection quality can be observed, especially in the Se and in recordings later during pregnancy. This might be an effect of the characteristics of the fetal movements or the ECG signal used to detect them changing with GA. Independent of the exact source of the GA-based effect, we can verify in Table 6.5 that training the model on two separate datasets split by GA markedly improves the detection results. Further improvements might be obtained by splitting the dataset in a more fine-grained manner to get GA-specific feature weights, although care has to be taken that sufficient representative data remains in each part of the dataset for both training and validation.

Further improvement of the ExG-based fetal movement detection might be possible by including additional features and use a machine-learning approach similar to the accelerometer-based method to perform feature selection. Another clear improvement option consists in using features extracted from multiple ExG leads, as currently only a single fixed lead is used. Using multiple leads might improve maternal ECG removal, the accuracy of fetal QRS detection, and improve the quantity and quality of the information contained in the extracted features. Multi-lead

Table 6.5 Fetal body movement detection quality using changes in the fetal QRS shape observed from a single vertical bi-polar abdominal ExG lead for a model trained for each patient, for all patients in a GA group (GA), and for all patients in the dataset (all)

GA	Patient specific			Cross validation (GA)			Cross validation (all)		
	Se	Sp	Ac	Se	Sp	Ac	Se	Sp	Ac
>35 weeks	0.78	0.70	0.70	0.74	0.68	0.69	0.65	0.63	0.57
<35 weeks	0.58	0.70	0.67	0.56	0.70	0.64	0.53	0.70	0.65
All	0.64	0.70	0.68	0.62	0.69	0.67	0.56	0.68	0.63

feature extraction might, in addition to fetal movement detection, also offer the possibility to determine the position and orientation of the fetus using, for example, vectorcardiographic loop alignment [35].

Finally, the fetal ECG-based fetal motion detection method might be used in conjunction with the previously presented accelerometer-based method to provide a complete picture of fetal motility. This would allow to move from movement counting to tracking of movement patterns and changes in the type and amount of movements, providing deeper insight into the fetal health.

6.4   Conclusion

New digital health technologies for fetal health monitoring are becoming available in obstetric care, enabling an improvement in the quality of care by improving the quality of measured fetal health features and increasing patient comfort. At the same time, a second wave of improvements in measurement technology is underway which promises to enable unobtrusive continuous fetal monitoring from the comfort of the home. This trend might provide a number of key benefits for all parties involved. Continuous monitoring will result in improved pregnancy outcomes by enabling timely intervention in case of any adverse events. At-home monitoring will improve patient comfort, while reducing the strain on the healthcare system by moving patients out of the hospital. In addition, these new techniques allow putting more information in the hands of pregnant women, offering an opportunity to reduce anxiety and allow them to make choices during pregnancy without second-guessing. However, care has to be taken to avoid over-information while addressing the women's needs and delivering clinical value to improve care. Finally, the large amount of previously unavailable longitudinal data might accelerate scientific and clinical research and lead to new insights in fetal health.

References

1. Penner, S.: Following the Trend from Inpatient to Outpatient Care. Springer: New York (2013)
2. Buysse, H., De Moor, G., Van Maele, G., Baert, E., Thienpont, G., Temmerman, M.: Cost-effectiveness of telemonitoring for high-risk pregnant women. Int. J. Med. Inform. 77(7), 470–476 (2008)
3. Haluza, D., Jungwirth, D.: ICT and the future of healthcare: aspects of pervasive health monitoring. Inform. Health Soc. Care. 43(1), 1–11 (2018)
4. Sheikh, M., Hantoushzadeh, S., Shariat, M.: Maternal perception of decreased fetal movements from maternal and fetal perspectives, a cohort study. BMC Pregnancy Childbirth. 14(1), 286 (2014)
5. Crawford, A., Hayes, D., Johnstone, E.D., Heazell, A.E.P.: Women's experiences of continuous fetal monitoring – a mixed-methods systematic review. Acta Obstet. Gyn. Scan. 96(12), 1404–1413 (2017)

6. Schramm, K., Lapert, F., Nees, J., Lempersz, C., Oei, S.G., Haun, M.W., Maatouk, I., Bruckner, T., Sohn, C., Schott, S.: Acceptance of a new non-invasive fetal monitoring system and attitude for telemedicine approaches in obstetrics: a case–control study. Arch. Gynecol. Obstet. **298**(6), 1085–1093 (2018)

7. Goldenberg, R.L., Culhane, J.F., Iams, J.D., Romero, R.: Epidemiology and causes of preterm birth. Lancet. **371**(9606), 75–84 (2008)

8. Alves, D.S., Times, V.C., da Silva, É.M.A., Melo, P.S.A., de Araújo Novaes, M.: Advances in obstetric telemonitoring: A systematic review. Int. J. Med. Inform. **134**, 104004 (2019)

9. HITC: GE healthcare acquires fetal monitoring technology Monica healthcare to expand digital maternal-infant care footprint, 2017

10. Jonathan Shieber. Electronics Giant Philips Invests in Monitoring an Infomation Platform for Expecting Mothers, 2019

11. Gibb, D., Arulkumaran, S.: Fetal Monitoring in Practice. Elsevier Health Sciences (2017)

12. Hasan, M.A., Ibrahimy, M.I., Reaz, M.B.I.A.N.D.: An efficient method for fetal electrocardiogram extraction from the abdominal electrocardiogram signal. J. Comput. Sci. **9**(5), 619–623 (2009)

13. Heazell, A.E.P., Frøen, J.F.: Methods of fetal movement counting and the detection of fetal compromise. J. Obstet. Gynaecol. **28**(2), 147–154 (2008)

14. Jezewski, J., Wrobel, J., Horoba, K.: Comparison of doppler ultrasound and direct electrocardiography acquisition techniques for quantification of fetal heart rate variability. IEEE Trans. Biomed. Eng. **53**(5), 855–864 (2006)

15. Besinger, R.E., Johnson, T.R.B.: Doppler recordings of fetal movement: Clinical correlation with real-time ultrasound. Obstet. Gynecol. **74**(2), 277–280 (1989)

16. DiPietro, J.A., Costigan, K.A., Pressman, E.K.: Fetal movement detection: comparison of the toitu actograph with ultrasound from 20 weeks gestation. J. Matern. Fetal Neonatal. Med. **8**(6), 237–242 (1999)

17. Peters, M., Crowe, J., Piéri, J.-F., Quartero, H., Hayes-Gill, B., James, D., Stinstra, J., Shakespeare, S.: Monitoring the fetal heart non-invasively: a review of methods. J. Perinat. Med. **29**(5), 408–416 (2001)

18. Barnett, S.B., Maulik, D.: Guidelines and recommendations for safe use of doppler ultrasound in perinatal applications. J. Matern. Fetal Med. **10**(2), 75–84 (2001)

19. Cremer, M.: Über die direkte ableitung der aktionsströme des menschlichen herzens vom oesophagus und über das elektrokardogramm des fötus. Lehmann, 1906

20. Burkman, R.T.: Williams Obstetrics, vol. 24, 24th edn. American Medical Association (2010)

21. Vullings, R., Verdurmen, K.M.J., Hulsenboom, A.D.J., Scheffer, S., de Lau, H., Kwee, A., Wijn, P.F.F., Amer-Wåhlin, I., van Laar, J.O.E.H., Oei, S.G.: The electrical heart axis and st events in fetal monitoring: A post-hoc analysis following a multicentre randomised controlled trial. Plos One. **12**(4), 1–11 (2017)

22. van Leeuwen, P., Halier, B., Bader, W., Geissler, J., Trowitzsch, E., Grönemeyer, D.H.W.: Magnetocardiography in the diagnosis of fetal arrhythmia. BJOG Int. J. Obstet. Gynaecol. **106**(11), 1200–1208 (1999)

23. Vullings, R.: Non-invasive fetal electrocardiogram: analysis and interpretation. PhD thesis, Doctoral Dissertation (2010)

24. Tiwari, A.K., Chourasia, V.: A review and comparative analysis of recent advancements in fetal monitoring techniques. Crit. Rev. Biomed. Eng. **36**(5–6), 335–373 (2008)

25. Altini, M., Mullan, P., Rooijakkers, M., Gradl, S., Penders, J., Geusens, N., Grieten, L., Eskofier, B.: Detection of fetal kicks using body-worn accelerometers during pregnancy: trade-offs between sensors number and positioning. In: Engineering in Medicine and Biology Society (EMBC), 2016 IEEE 38th Annual International Conference of the, pp. 5319–5322. IEEE, 2016

26. Surlea, C., Kurugollu, F., Milligan, P., Ong, S.: Foetal motion classification using optical flow displacement histograms. In: Proceedings of the 4th International Symposium on Applied

Sciences in Biomedical and Communication Technologies, ISABEL '11, pp. 156:1–156:5, New York, 2011. ACM

27. Kovács, F., Horváth, C.: Ádám T Balogh, and Gábor Hosszú. Fetal phonocardiography—past and future possibilities. Comput. Methods Prog. Biomed. **104**(1), 19–25 (2011)

28. DiPietro, J.A., Hodgson, D.M., Costigan, K.A., Hilton, S.C., Johnson, T.R.B.: Development of fetal movement - fetal heart rate coupling from 20 weeks through term. Early Hum. Dev. **44**(2), 139–151 (1996)

29. Rooijakkers, M., Rabotti, C., de Lau, H., Oei, S., Bergmans, J., Mischi, M.: Feasibility study of a new method for low-complexity fetal movement detection from abdominal ECG recordings. IEEE J Biomed Health Inform. **20**, 1361–1368 (2016)

30. Biglari, H., Sameni, R.: Fetal motion estimation from noninvasive cardiac signal recordings. Physiol. Meas. **37**(11), 2003–2023 (2016)

31. Abdulhay, E.W., Oweis, R.J., Alhaddad, A.M., Sublaban, F.N., Radwan, M.A., Almasaeed, H.M.: Review article: Non-invasive fetal heart rate monitoring techniques. Biomed. Sci. **2**(3), 53–67 (2014)

32. U.S. Food & Drug Administration. Avoid fetal "keepsake" images, heartbeat monitors, 2014

33. Altini, M., Rossetti, E., Rooijakkers, M., Penders, J., Lanssens, D., Grieten, L., Gyselaers, W.: Variable-length accelerometer features and electromyography to improve accuracy of fetal kicks detection during pregnancy using a single wearable device. In: 2017 IEEE EMBS International Conference on Biomedical & Health Informatics (BHI), pp. 221–224. IEEE, 2017

34. Crowe, J.A., James, D., Hayes-Gill, B.R., Barratt, C.W., Pieri, J.-F.. Fetal surveillance, May 2005

35. Vullings, R., Mischi, M., Oei, S.G., Bergmans, J.W.M.: Novel Bayesian vectorcardiographic loop alignment for improved monitoring of ECG and fetal movement. I.E.E.E. Trans. Biomed. Eng. **60**(6), 1580–1588 (2013)

36. Oostendorp, T.F.: Modelling the Fetal ECG. Katholieke Universiteit te Nijmegen, 1989

37. Rooijakkers, M.J., Rabotti, C., Oei, S.G., Mischi, M.: Low-complexity R-peak detection for ambulatory fetal monitoring. Physiol. Meas. **33**(7), 1135–1150 (2012)

38. Zhong, W., Liao, L., Guo, X., Wang, G.: A deep learning approach for fetal qrs complex detection. Physiol. Meas. **39**(4), 045004 (2018)

39. Frederik Frøen, J., Heazell, A.E.P., Tveit, J.V.H., Saastad, E., Fretts, R.C., Flenady, V.: Fetal movement assessment. Semin. Perinatol. **32**(4), 243–246 (2008). Antenatal Testing: A Re-Evaluation

Chapter 7
T-Wave Alternans Identification in Direct and Indirect Fetal Electrocardiography

Laura Burattini, Ilaria Marcantoni, Amnah Nasim, Luca Burattini, Micaela Morettini, and Agnese Sbrollini

Contents

7.1   What Is T-Wave Alternans?

The electrocardiogram (ECG) is the recording of the heart electrical activity. In order to acquire this signal, clinical noninvasive electrocardiography implies the placement of electrodes on the skin of chest and limbs according to a conventional configuration that ensures standardization of the acquisition leads. By its nature, the ECG is a pseudo-periodic signal, each period of which represents a cardiac cycle. Three main waveforms are identified: the P wave, which represents the depolarization of atria; the QRS complex, which represents the depolarization of ventricles; and the T wave and U wave, if present, which represent the repolarization of ventricles. Anomalies affecting the T wave often indicate severe cardiovascular diseases and heart instability since reflecting ventricles abnormalities, which are recognized among the most critical conditions for the heart. Specifically, anomalies

L. Burattini (✉) · I. Marcantoni · A. Nasim · M. Morettini · A. Sbrollini
Department of Information Engineering, Università Politecnica delle Marche, Ancona, Italy
e-mail: l.burattini@univpm.it

L. Burattini
Department of Clinical Sciences, Ospedali Riuniti di Ancona (Salesi Hospital), Ancona, Italy

© Springer Nature Switzerland AG 2021
D. Pani et al. (eds.), *Innovative Technologies and Signal Processing in Perinatal Medicine*, https://doi.org/10.1007/978-3-030-54403-4_7

concerning ventricular repolarization showed to be strongly correlated to life-threatening ventricular tachyarrhythmias and high risk of sudden cardiac death at any age.

T-wave alternans (TWA, depicted in Fig. 7.1) is an electrophysiological phenomenon that may affect the ECG: it consists in a beat-to-beat fluctuation of the morphology (amplitude, shape, and/or polarity) of the electrocardiographic T wave occurring at a stable heart rate (HR). TWA is non-stationary [1] (since its amplitude and duration generally vary with time) and lead-dependent [2] (since its amplitude depends on the ECG leads). It originates from anomalies in the timing of repolarization at the level of the ventricle cells. Traditionally, two classes of TWA are considered: the macroscopic TWA, visually noticeable at standard display scales; and the microscopic TWA, not visually noticeable at standard display scales.

For the first time observed by Hering in 1908, macroscopic TWA was initially considered as a curiosity of the ECG morphology. In 1975 Schwartz and Malliani discovered its association with the long QT syndrome, and in 1984 Adam et al. showed the existence of microscopic TWA. Unlike macroscopic TWA, microscopic TWA requires specifically designed automatic methods to be detected and quantified [1, 3, 4].

TWA importance is grounded on its recognized role as a noninvasive risk marker. Specifically, many experimental and clinical studies indicate TWA as a reliable risk index of severe cardiac arrhythmic events, especially life-threatening arrhythmias and sudden cardiac death. Certainly, macroscopic TWA is a marker of severe heart instability, but also microscopic TWA is clinically significant and deserves even more attention because more frequently observed. TWA is known to be HR-dependent and is often visible only at high HR. Moreover, TWA identification requires HR stability, otherwise the changes in the T-wave amplitude may be HR-driven. Therefore, in many studies presented in the literature, TWA is usually detected under pacing conditions or during exercise, when HR is higher and less affected by heart-rate variability (HRV). In these studies, TWA showed to be a promising risk stratifier for ventricular arrhythmic events in several pathological heart conditions, like ischemic and non-ischemic cardiomyopathy, long QT syndrome, myocardial infarction, congestive heart failure and coronary heart disease [5, 6]. Some studies also reported a certain level of TWA in healthy subjects, even if always lower than in patients, suggesting the existence of physiological levels [5, 7, 8]. Nevertheless, the universal definition of threshold values at the verge of abnormal conditions is difficult to be assessed because they depend on automatic methods used to detect TWA [8].

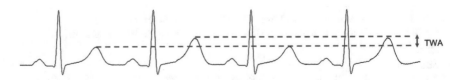

Fig. 7.1 Example of an ECG affected by macroscopic TWA

7.2 Direct and Indirect Fetal Electrocardiography

Fetal electrocardiography is the recording of the electrical activity of the fetal heart. Morphologically, the fetal ECG (FECG) shows the standard P-QRS-TU pattern also present in adult ECG [9] but with different HR values. Specifically, fetal HR, typically around 120–160 bpm in normal conditions [10], is two to three times higher than adult HR [11].

There are mainly two techniques currently used for FECG acquisition: the direct/internal one (Fig. 7.2, upper panel) and the indirect/external one (Fig. 7.2, lower panel) [12]. The direct fetal electrocardiography consists in the placement of a spiral wire electrode on the fetal scalp to directly record the FECG (DI_FECG). It is applicable only during the delivery, when rupture of the chorioamniotic membrane has occurred and there is enough dilatation of the uterine cervix for the electrode placement [9]. This technique is invasive and may provoke the risk of infection for the mother, and a mark or a small cut on the fetus's head. Thus, it is recommended only in cases of a risky pregnancy [12]. DI-FECG signal quality is high: it has a high amplitude (about hundreds of µV) with a good signal-to-noise ratio (SNR), making the signal processing easy.

The indirect fetal electrocardiography consists in the placement of surface electrodes on the mother's abdomen to indirectly record the FECG (IN-FECG). Specifically, surface electrodes are placed around the navel of the mother and a reference electrode is placed above the symphysis pubica. It is applicable in early pregnancy or during the late pregnancy, when the vernix caseosa (an anatomical layer that electrically shields the fetus), the progressive formation of which produces a thick layer around 30 weeks of gestation, slowly dissolves [9]. Indirect fetal electrocardiography is a nonstress and noninvasive test and no risks for both mother and fetus have been reported. IN-FECG quality is low: it has a low amplitude (about

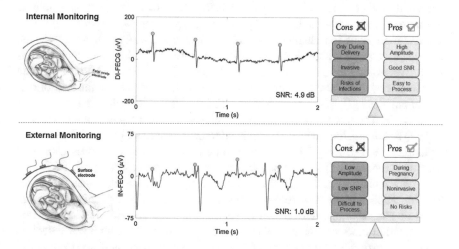

Fig. 7.2 Internal (upper panel) and external (lower panel) fetal electrocardiography

tens of μV) and is affected by a high level of noise, often hiding the signal of interest. Indeed, the surface electrodes also record the maternal ECG, maternal and fetal muscular noise, and other kinds of maternal internal noises [10, 12, 13]. This results in low SNR and the need of complex signal processing procedures.

7.3   T-Wave Alternans in Fetal Electrocardiography

7.3.1   Why Measuring T-Wave Alternans in Fetal Electrocardiography?

Health and well-being of a nation is often estimated through infant mortality, since factors inducing infant mortality have an impact on the health of the general population [14]. Infant death is often related to genetic irregularities in the ion-channel function elected to regulate cardiac repolarization [15]. It is significantly important to have a risk index that could be used to highlight possible severe and dangerous abnormalities of the cardiac repolarization in order to take action in time. One of the risk markers used for the adult population is TWA. TWA could be exploited to predict and prevent malignant ventricular repolarization irregularities also in the fetal population in order to decrease infant deaths in the future. However, TWA is much more difficult to be detected in fetuses than in adults; it manifests spontaneously, with variable duration and occurrence rate, even if it can occur continually in severe situations. Knowledge on the etiology of TWA in utero is very limited [15, 16].

A deeper evaluation of electrophysiology of repolarization characteristics before birth is important because it may make possible to evaluate the electrophysiology of fetuses at higher risk for sudden death [17, 18]. Fetuses with repolarization abnormalities frequently go to suboptimal outcome, infant death, or in utero death. The hormonal mother state during pregnancy differently affects the electrophysiologic substrate of both mother and fetus. Specifically, high levels of estrogens can stabilize maternal heart ion channels but not the fetal ones. Moreover, hemodynamic changes concurrent with arrhythmias or heart failure can be due to anomalous oxygen arrival at the immature myocardium. Both the timing and the level of hypoxia may influence the fetal electrophysiological substrate, especially in the case of cardiac anomalies [17]. Sudden death incidence is higher in utero than in other phases of life cycle. The reason for this is almost unknown, even if several studies found it in the QT prolongation, which is correlated to increased risk of ventricular tachyarrhythmias. Indeed, long QT was found to be more frequent in fetuses with poor outcomes; moreover, it is often associated with concomitant TWA [15]. TWA possibly puts fetuses at high risk of ventricular tachycardia and its detection makes genetic ion channelopathies research necessary, as well as investigation of mother levels of magnesium, calcium, and D vitamin [16]. An accurate investigation on repolarization-function quality, which can be performed through TWA detection and quantification, could allow an effective in-utero pharmacological treatment to restore regular cardiac activity and avoid premature delivery and so allow well-timed neonatal care. The possibility of assessing repolarization prenatally adds value and reliability to fetal genetic screening and could prove lifesaving [15, 18]. Therefore, being the possible cause of some unexplained fetal deaths, fetal TWA investigation might become the right approach to solve many currently unexplained fetal problems [15].

7.3.2   Improved Fetal Pan-Tompkins Algorithm for Automatic Detection of Fetal R Peaks

Any fetal TWA investigation starts with automatic identification of fetal R peaks and FECG filtering, followed by the application of a proper algorithm for TWA detection. The Pan-Tompkins algorithm (PTA, upper panel of Fig. 7.3) [19] is a well-known algorithm for adult R-peak identification. Briefly, it is composed of four steps that are a bandpass filtering step (cut-off frequencies of 5 Hz and 15 Hz), a 25 ms differentiation step, a squaring operation step, and a 150 ms moving-window integration step. To correctly identify the R-peak positions, two adaptive thresholds (Sf and Si) are considered to validate the local maxima detected from the filtered ECG (filtECG, obtained after the first step) and the integrated ECG (intECG, obtained after the fourth step) signals. If a local maximum is present in both filtECG and intECG signals, it is confirmed as R peak; otherwise it is rejected.

In order to use the PTA algorithm for fetal R-peak identification, PTA was recently adapted to FECG features [20]. The adapted algorithm is called the improved fetal Pan-Tompkins algorithm (IFPTA, lower panel of Fig. 7.3). By considering that FECG has the same morphology of adult ECG but is characterized by higher (about double) HR, the first and the fourth steps of PTA needed adaptation. Specifically, the cut-off frequencies of the bandpass filtering step become 9 Hz and 27 Hz, and the moving-window integration step is performed over an 80 ms window. Finally, in order to remove the false-positive and false-negative peaks, a fetal R-peak corrector is added in cascade. The corrector extracts a 9-beat window around each selected R-peak position and computes the mean RR interval and the mean fetal QRS complex in the window. If the RR interval associated with the selected R-peak position is significantly shorter or longer than the mean RR interval, or the correlation between the QRS complex associated with the selected R-peak position and the mean QRS complex is weak, the selected R-peak position is corrected. Details about the fetal R-peak corrector can be found in [20]. Briefly, false-positive detections, identified by short RR intervals and possibly low correlations, are eliminated; instead, false-negative detections, identified by long RR interval, are added.

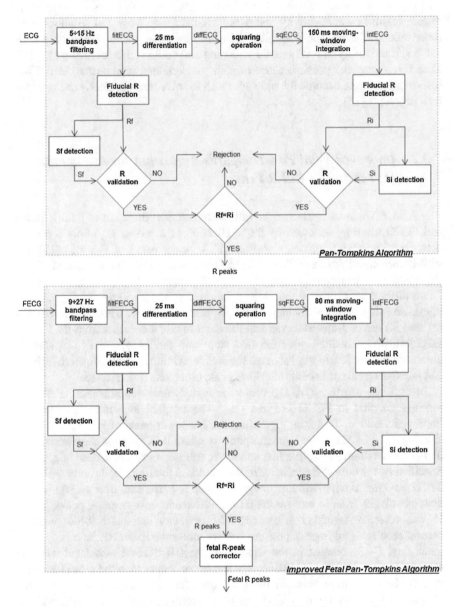

Fig. 7.3 Block diagrams of the PTA (upper panel) and of the IFPTA (lower panel)

7.3.3 Segmented-Beat Modulation Method for Electrocardiographic Filtering

The segmented-beat modulation method (SBMM) [21, 22] is a template-based noise cancellation procedure aimed at increasing the clinical utility of an ECG affected by high levels of noise. The block diagram of the SBMM algorithm is depicted in Fig. 7.4. The SBMM considers a noisy ECG and its corresponding R-peak sequence as inputs and provides a clean ECG as output. The input R peaks are used to identify all cardiac cycles (CCs): each CC is defined as the noisy ECG segment included between Δ ms before an R peak and Δ ms before the next one. All CCs are then segmented into QRS segments (from the beginning of each CC to Δ ms after its R peak) and TUP segments (Δ ms after the R peak to the end of each CC). Duration of cardiac ventricular depolarization (QRS segments) is assumed to be constant ($2 \cdot \Delta$ ms), while the rest of CCs (TUP segments) are assumed to be RR-dependent (TUP = RR-$2 \cdot \Delta$). Thus, after CC segmentation, a modulation (compression/stretch) of TUP segments is performed to match the median TUP segment, directly computed from the median RR. All QRS and TUP segments are then reconstructed in order to obtain all equally long CCs. A median operation over all CCs is performed to compute a template beat: thanks to the median operation properties, the template beat appears denoised. The template beat is then segmented into median QRS segment and median TUP segment. The median TUP segment (repeated for each beat) is demodulated (stretch/compression) back to the original TUP segment lengths. Lastly, the median QRS segment (repeated for each beat) and the demodulated TUP segments are concatenated to construct the demodulated CCs, which are in turn concatenated to obtain the clean ECG. Thanks to the demodulation procedure, the clean ECG maintains the same HRV properties of the original noisy ECG.

7.3.4 Heart-Rate Adaptive Match Filter Method

The detection of microvolt TWA can be performed only through automatic methods. One of the methods that can be used to investigate microvolt TWA is the heart-rate adaptive match filter (HRAMF) [1]. The HRAMF-based procedure for TWA identification and quantification is described through the block diagram in Fig. 7.5.

Before the application of the method, a preprocessing of ECG is needed. Initially, ECG windows are recursively extracted [15] (window features are application-dependent). Each window is low-pass filtered (cut-off frequency: 35 Hz) to remove high frequency. The filter is a sixth-order bidirectional Butterworth filter that prevents filtering delay and oscillation in the filtering passing band. Considering the related R peaks, subtraction of baseline, computed as a third-order spline interpolation, is performed. The further analysis takes into consideration the extracted window, set as the part of the ECG window that includes the M central heartbeats (numerical value of M is application-dependent). These beats are used to compute the mean RR interval, the RR standard deviation, and the TWA frequency (F_{TWA}); which is defined as half of mean RR interval reciprocal. Moreover, in each extracted window the median beat is estimated and its correlation with each present beat is evaluated: if correlation is weak, the beat is classified as ectopic or noisy and so replaced by the median one [15].

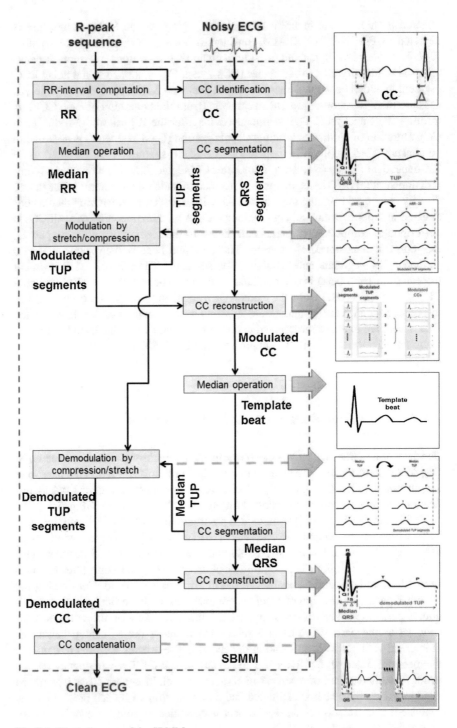

Fig. 7.4 Block diagram of the SBMM

Fig. 7.5 Block diagram of HRAMF-based procedure for TWA identification and quantification

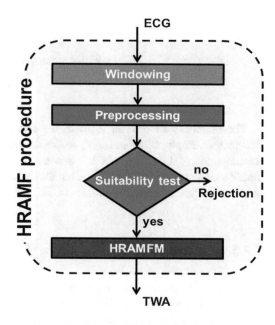

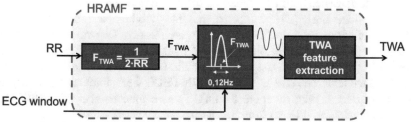

Fig. 7.6 Block diagram of the HRAMF

After preprocessing, the extracted window has to be tested in terms of suitability for TWA detection. Particularly, if RR standard deviation does not overcome 10% of mean RR interval and replaced heartbeats are less than 10% of M, the extracted window goes to analysis of HRAMF; if not, it is rejected [15]. The HRAMF, the block diagram of which is depicted in Fig. 7.6, exploits F_{TWA} to design a bandpass filter with a central frequency corresponding to F_{TWA} and a passing band that is very narrow (0.12 Hz-wide).

The limited width of the filter is able to respect physiological HRV (computed as RR standard deviation) and, in the same time, to maintain frequency band pertaining only to TWA phenomenon. Specifically, the filter is designed as a cascade of a sixth-order bidirectional Butterworth low-pass filter (LPF; cut-off frequency $f_{LPF} = F_{TWA} + 0.06$ Hz) and a sixth-order bidirectional Butterworth high-pass filter (HPF; cut-off frequency $f_{HPF} = F_{TWA} - 0.06$ Hz). Its squared module is given by the following equation [5]:

$$\left|H_{\text{HRAMF}}\left(\omega\right)\right|^2 = \left|H_{\text{LPF}}\left(\omega\right)\right|^2 \cdot \left|H_{\text{HPF}}\left(\omega\right)\right|^2 = \frac{1}{1+\left(\dfrac{\omega}{\omega_{LPF}}\right)^{2n}} \cdot \frac{\left(\dfrac{\omega}{\omega_{HPF}}\right)^{2n}}{1+\left(\dfrac{\omega}{\omega_{HPF}}\right)^{2n}}. \quad (7.1)$$

The extracted window in input to the HRAMF undergoes filtering and then becomes an amplitude modulated pseudo-sinusoidal signal. If TWA is present, the pseudo-sinusoidal signal has its maxima and minima over the T waves and twice its amplitude is quantification of TWA amplitude. If TWA is not present, the pseudo-sinusoidal signal is actually a constant and TWA amplitude is zero [15].

7.3.5 A Clinical Study

The reported clinical study presents an example of fetal TWA extraction from simultaneously acquired DI-FECG and IN-FECG.

Data Data consisted in FECG recordings, available from the open source Physionet (www.physionet.org) [23], collected in the "Abdominal and Direct Fetal Electrocardiogram Database" [24] by the research team of Department of Obstetrics at the Medical University of Silesia. FECG signals were acquired from 5 pregnant women (between 38th and 41st week of gestation) by internal and external monitoring. For each fetus, one DI-FECG and four IN-FECG were simultaneously recorded (sampling rate: 1 kHz; duration: 5 min). Thus, a total number of 25 FECG (5 DI-FECG and 20 IN-FECG) from 5 fetuses were collected.

Methods Figure 7.7 depicts the procedures for TWA extraction from DI-FECG (Fig. 7.7A) and IN-FECG (Fig. 7.7B) [15, 25]. Firstly, TWA was detected by the HRAMF-based procedure (window length: 35 s; moving step for window extraction: 1 s; M: 32) from DI-FECG, considering its fetal R peaks extracted by IFPTA [20]. Secondly, considering the high level of noise, IN-FECG was denoised from maternal ECG (MECG) interference before TWA extraction. SBMM was applied in order to obtain MECG, after the maternal R-peaks identification by standard PTA [19]. By subtraction, MECG was removed from the original recording, obtaining a denoised IN-FECG. By considering the difference in amplitude in relation to DI-FECG, IN-FECG was amplified by a 4.8 scale factor [25]. Finally, TWA was detected by HRAMF procedure (window length: 35 s; moving step for window extraction: 1 s; M: 32) from denoised IN-FECG, considering its fetal R peaks extracted by IFPTA [20].

Quality of FECG was quantified in term of SNR, computed as:

Fig. 7.7 Procedure for
TWA extraction from
DI-FECG (panel **A**) and
from IN-FECG (panel **B**)

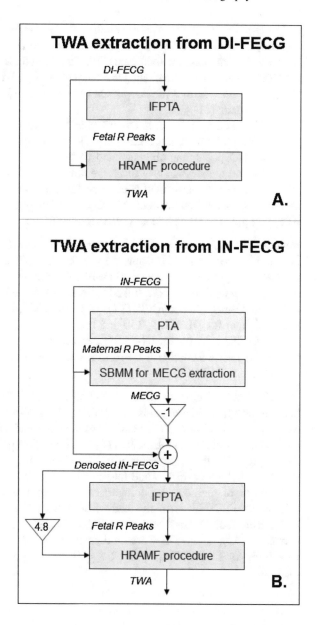

$$SNR = 20 \cdot \log_{10} \left(\frac{FECG_{Amplitude}}{FNOISE_{Amplitude}} \right) \qquad (7.2)$$

where $FECG_{Amplitude}$ was obtained as the mean peak-to-peak amplitude over beats
and $FNOISE_{Amplitude}$ was obtained as four times standard deviation of the fetal noise.
FECG and FNOISE were separated by SBMM [13].

For each FECG, the number of suitable windows (NW) was considered. For each window, FECG features and TWA features were extracted. Considered FECG features are the fetal HR, the fetal mean RR interval and the fetal HRV, while considered TWA features are the mean TWA frequency (F_{TWA}), the mean TWA amplitude (M_{TWA}), the maximum TWA amplitude (MAX_{TWA}), and the standard deviation of TWA amplitude (SD_{TWA}).

In order to compare TWA in DI-FECG and IN-FECG, only one lead of IN-FECG was selected. The selection criteria are: I) the selected lead must have the highest value of MAX_{TWA} among all leads; and II) the selected lead must have NW higher than 15% and higher than NW of all the other leads. The comparison of DI-FECG and IN-FECG was performed computing the Pearson's correlation coefficient (ρ) and the regression line between MAX_{TWA} and HR (statistical significance level at 0.05) computed in DI-FECG and selected IN-FECG.

Results FECG features and TWA features were extracted from all FECG recordings and they are reported in Table 7.1 [15, 25]. Overall, all DI-FECG were suitable for TWA analysis, while 9 out of 20 IN-FECG were rejected. The suitability assessment can be quantified by NW, linked to SNR. NW ($69 \pm 29\%$) and SNR (-1 ± 8 dB) of DI-FECG are higher than NW ($17 \pm 18\%$) and SNR (-4 ± 5 dB) of IN-FECG. HR (129 ± 3 bpm for DI-FECG; 130 ± 5 bpm for IN-FECG), RR (467 ± 10 ms for DI-FECG; 461 ± 17 ms for IN-FECG), and consequently F_{TWA} (1.07 ± 0.02 Hz for DI-FECG; 1.09 ± 0.04 Hz for IN-FECG) were homogenous among fetuses in both DI-FECG and IN-FECG.

TWA features computed from DI-FECG and IN-FECG have the same order of magnitude: M_{TWA} distribution is 9 ± 2 µV for DI-FECG and 11 ± 5 µV for IN-FECG; MAX_{TWA} distribution is 30 ± 11 µV for DI-FECG and 21 ± 12 µV for IN-FECG; and, SD_{TWA} distribution is 6 ± 2 µV for DI-FECG and 7 ± 3 µV for IN-FECG. Moreover, SD_{TWA} presented the same order of magnitude as M_{TWA}, highlighting a great TWA variability in both DI-FECG and IN-FECG.

In order to perform the comparison between DI-FECG and IN-FECG, the selected IN-FECG leads are the lead 4 from fetus 1, the lead 4 from fetus 4, and the lead 1 from fetus 5 (leads with 'a' in Table 7.1). Pearson's correlation coefficients between MAX_{TWA} and HR were moderate for DI-FECG ($\rho = 0.64$; $P = 0.24$) and very high for IN-FECG ($\rho = 0.99$; $P = 0.02$). Regression lines are depicted in Fig. 7.8.

7.4   Final Remarks

The identification and quantification of TWA was successful in both DI-FECG and IN-FECG. Even if the first one is more suitable for TWA detection, because less affected by interferences, it is in the same time almost not practicable in clinical practice, because of its invasiveness. Moreover, the IN-FECG has to deal with the

Table 7.1 FECG features and TWA features extracted from all FECG recording

Fetus/lead	SNR (dB)	NW	HR (bpm)	RR (ms)	HRV (ms)	F_{TWA} (Hz)	M_{TWA} (µV)	MAX_{TWA} (µV)	SD_{TWA} (µV)
F1/DI-FECG	5	242(91.7%)	129	466	14	1.07	6	20	4
F1/IN-FECG1	−1	2(0.8%)	128	468	0	1.07	6	7	3
F1/IN-FECG2	−3	0(0%)	–	–	–	–	–	–	–
F1/IN-FECG3	1	44(16.7%)	127	471	5	1.06	6	10	4
F1/IN-FECG4[a]	3	71(26.9%)	127	472	6	1.06	6	24	5
F2/DI-FECG	−3	173(65.5%)	129	466	23	1.08	8	29	6
F2/IN-FECG1	−15	0(0%)	–	–	–	–	–	–	–
F2/IN-FECG2	−8	0(0%)	–	–	–	–	–	–	–
F2/IN-FECG3	−9	0(0%)	–	–	–	–	–	–	–
F2/IN-FECG4	−7	0(0%)	–	–	–	–	–	–	–
F3/DI-FECG	−4	52(19.7%)	124	483	11	1.04	8	19	5
F3/IN-FECG1	−16	0(0%)	–	–	–	–	–	–	–
F3/IN-FECG2	−7	0(0%)	–	–	–	–	–	–	–
F3/IN-FECG3	−8	0(0%)	–	–	–	–	–	–	–
F3/IN-FECG4	−6	10(3.8%)	126	476	1	1.05	23	34	10
F4/DI-FECG	9	236(89.4%)	131	459	15	1.09	9	34	8
F4/IN-FECG1	−2	2(0.8%)	128	468	1	1.07	5	6	4
F4/IN-FECG2	−5	1(0.4%)	144	416	0	1.20	14	14	9
F4/IN-FECG3	1	45(17.0%)	132	456	17	1.10	8	19	5
F4/IN-FECG4[a]	4	107(40.5%)	130	461	16	1.09	12	33	10
F5/DI-FECG	−12	202(76.5%)	130	460	23	1.09	12	47	8

(continued)

Table 7.1 (continued)

Fetus/lead	SNR (dB)	NW	HR (bpm)	RR (ms)	HRV (ms)	F_{TWA} (Hz)	M_{TWA} (μV)	MAX_{TWA} (μV)	SD_{TWA} (μV)
F5/IN-FECG1[a]	1	138(52.3%)	133	452	21	1.11	12	41	9
F5/IN-FECG2	−3	74(28.0%)	131	458	14	1.09	11	32	7
F5/IN-FECG3	−4	0(0%)	–	–	–	–	–	–	–
F5/IN-FECG4	−1	3(1.1%)	127	474	1	1.06	14	15	8
TOT DI-FECG	*−1 ± 8*	*181 ± 77 (69 ± 29%)*	*129 ± 3*	*467 ± 10*	*17 ± 5*	*1.07 ± 0.02*	*9 ± 2*	*30 ± 11*	*6 ± 2*
TOT IN-FECG	*−4 ± 5*	*45 ± 48 (17 ± 18%)*	*130 ± 5*	*461 ± 17*	*7 ± 8*	*1.09 ± 0.04*	*11 ± 5*	*21 ± 12*	*7 ± 3*

[a]Selected IN-FECG leads for the comparison assessment between DI-FECG and IN-FECG

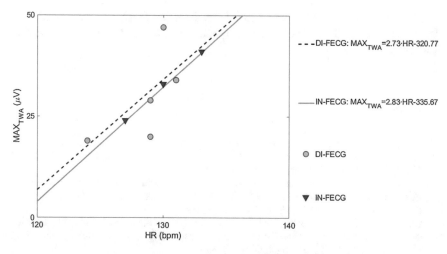

Fig. 7.8 Regression lines between MAXTWA and HR computed in DI-FECG (blue circles and dashed black line) and selected IN-FECG (red triangles and grey solid line)

dependency of TWA from acquisition lead, while the DI-FECG is unavoidably conditioned by the stressful status of the fetus during delivery [15].

According to our experience, HRAMF is a good method for TWA extraction. Thanks to its theoretical approach, HRAMF is robust against noises and interferences in many frequency bands, avoiding false positives; moreover, it allows a reliable interpretation of TWA, since it respects the non-stationary nature of the phenomenon. These advantages make HRAMF particularly suitable in the case of fetal electrocardiography, because FECG is more affected by noises and interferences and, if present, fetal TWA is highly variable in duration and amplitude [16].

The main finding from TWA analysis on fetal electrocardiography is that fetuses show TWA, even if they are healthy, while in adult applications TWA manifestation is typically associated with a pathological condition. Another aspect to keep in consideration is that FECG has a different amplitude with respect to adult ECG and this can influence the quantification of TWA; consequently, if a certain value of TWA is considered physiologic in healthy adults, the same value could assume a different meaning in case of fetuses. At the same time, values of TWA evaluated as high if referred to ECG amplitude may be affected by the high HR, which is typical in fetuses [15]. Still, threshold or reference values of TWA are difficult to be defined for both adults and fetuses, also because they depend on the used TWA identification method [26].

As reported in existing literature, TWA is strongly variable in time [15]. Moreover, comparable results about TWA evaluation and characterization were found when the HRAMF was applied on DI-FECG and IN-FECG, but it is necessary to consider an amplitude-correction factor so that amplitude of both kinds of acquisition is of the same order.

Finally, it is remarkable the fact that TWA detection is reliable also considering IN-FECG, even if it implies having many more artifacts and interferences to deal with [25].

References

1. Burattini, L., Zareba, W., Burattini, R.: Adaptive match filter based method for time vs. amplitude characterization of microvolt ECG T-wave alternans. Ann. Biomed. Eng. **36**, 1558–1564 (2008). https://doi.org/10.1007/s10439-008-9528-6
2. Burattini, L., Man, S., Burattini, R., Swenne, C.A.: Comparison of standard versus orthogonal ECG leads for T-wave alternans identification. Ann. Noninvasive Electrocardiol. **17**, 130–140 (2012). https://doi.org/10.1111/j.1542-474X.2012.00490.x
3. Burattini, L., Bini, S., Burattini, R.: Comparative analysis of methods for automatic detection and quantification of microvolt T-wave alternans. Med. Eng. Phys. **31**, 1290–1298 (2009). https://doi.org/10.1016/j.medengphy.2009.08.009
4. Burattini, L., Bini, S., Burattini, R.: Correlation method versus enhanced modified moving average method for automatic detection of T-wave alternans. Comput. Methods Prog. Biomed. **98**, 94–102 (2010). https://doi.org/10.1016/j.cmpb.2010.01.008
5. Burattini, L., Bini, S., Burattini, R.: Repolarization alternans heterogeneity in healthy subjects and acute myocardial infarction patients. Med. Eng. Phys. **34**, 305–312 (2012). https://doi.org/10.1016/j.medengphy.2011.07.019
6. Burattini, L., Man, S., Fioretti, S., et al.: Heart rate-dependent hysteresis of T-wave alternans in primary prevention ICD patients. Ann. Noninvasive Electrocardiol. **21**, 460–469 (2016). https://doi.org/10.1111/anec.12330
7. Burattini, L., Zareba, W., Burattini, R.: Identification of gender-related normality regions for T-wave alternans. Ann. Noninvasive Electrocardiol. **15**, 328–336 (2010). https://doi.org/10.1111/j.1542-474X.2010.00388.x
8. Burattini, L., Zareba, W., Burattini, R.: Assessment of physiological amplitude, duration, and magnitude of ECG T-wave alternans. Ann. Noninvasive Electrocardiol. **14**, 366–374 (2009). https://doi.org/10.1111/j.1542-474X.2009.00326.x
9. Agostinelli, A., Grillo, M., Biagini, A., et al.: Noninvasive fetal electrocardiography: an overview of the signal electrophysiological meaning, recording procedures, and processing techniques. Ann. Noninvasive Electrocardiol. **20**, 303–313 (2015). https://doi.org/10.1111/anec.12259
10. Sameni, R., Clifford, G.D.: A review of Fetal ECG signal processing issues and promising directions. Open Pacing Electrophysiol. Ther. J. **3**, 4–20 (2010). https://doi.org/10.2174/1876536X01003010004
11. Van Mieghem, T., DeKoninck, P., Steenhaut, P., Deprest, J.: Methods for prenatal assessment of fetal cardiac function. Prenat. Diagn. **29**, 1193–1203 (2009). https://doi.org/10.1002/pd.2379
12. Hasan, M.A., Reaz, M.B.I., Ibrahimy, M.I., et al.: Detection and processing techniques of FECG signal for fetal monitoring. Biol. Proced. Online. **11**, 263–295 (2009). https://doi.org/10.1007/s12575-009-9006-z
13. Agostinelli, A., Sbrollini, A., Burattini, L., et al.: Noninvasive fetal electrocardiography part II: segmented-beat modulation method for signal denoising. Open Biomed. Eng. J. **11**, 25–35 (2017). https://doi.org/10.2174/1874120701711010025
14. Parks, S.E., Erck Lambert, A.B., Shapiro-Mendoza, C.K.: Racial and ethnic trends in sudden unexpected infant deaths: United States, 1995–2013. Pediatrics. **139**, e20163844 (2017). https://doi.org/10.1542/peds.2016-3844

15. Marcantoni, I., Vagni, M., Agostinelli, A., et al.: T-wave Alternans identification in direct Fetal electrocardiography. Comput. Cardiol. **44**, 1–4 (2017). https://doi.org/10.22489/CinC.2017.219-085
16. Yu, S., Van Veen, B.D., Wakai, R.T.: Detection of T-wave alternans in fetal magnetocardiography using the generalized likelihood ratio test. I.E.E.E. Trans. Biomed. Eng. **60**, 2393–2400 (2013). https://doi.org/10.1109/TBME.2013.2256907
17. Cuneo, B.F., Strasburger, J.F., Wakai, R.T.: Magnetocardiography in the evaluation of fetuses at risk for sudden cardiac death before birth. J. Electrocardiol. **41**, 116.e1–116.e6 (2008). https://doi.org/10.1016/j.jelectrocard.2007.12.010
18. Cuneo, B.F., Strasburger, J.F., Yu, S., et al.: In utero diagnosis of long QT syndrome by magnetocardiography. Circulation. **128**, 2183–2191 (2013). https://doi.org/10.1161/CIRCULATIONAHA.113.004840
19. Pan, J., Tompkins, W.J.: A real-time QRS detection algorithm. I.E.E.E. Trans. Biomed. Eng. **BME-32**, 230–236 (1985). https://doi.org/10.1109/TBME.1985.325532
20. Agostinelli, A., Marcantoni, I., Moretti, E., et al.: Noninvasive fetal electrocardiography part I: Pan-Tompkins' algorithm adaptation to fetal R-peak identification. Open Biomed. Eng. J. **11**, 17–24 (2017). https://doi.org/10.2174/1874120701711010017
21. Agostinelli, A., Giuliani, C., Burattini, L.: Extracting a clean ECG from a noisy recording: a new method based on segmented-beat modulation. Comput. Cardiol. **41**, 49–52 (2014)
22. Agostinelli, A., Sbrollini, A., Giuliani, C., et al.: Segmented beat modulation method for electrocardiogram estimation from noisy recordings. Med. Eng. Phys. **38**, 560–568 (2016). https://doi.org/10.1016/j.medengphy.2016.03.011
23. Goldberger, A.L., Amaral, L.A.N., Glass, L., et al.: PhysioBank, PhysioToolkit, and PhysioNet. Circulation. **101**, e215–e220 (2000). https://doi.org/10.1161/01.cir.101.23.e215
24. Jezewski, J., Matonia, A., Kupka, T., et al.: Determination of fetal heart rate from abdominal signals: evaluation of beat-to-beat accuracy in relation to the direct fetal electrocardiogram. Biomed. Tech. Eng. **57**, 383–394 (2012). https://doi.org/10.1515/bmt-2011-0130
25. Marcantoni, I., Sbrollini, A., Burattini, L., et al.: Automatic T-wave alternans identification in indirect and direct fetal electrocardiography. In: 2018 40th Annual International Conference of the IEEE Engineering in Medicine and Biology Society (EMBC), pp. 4852–4855. IEEE, Piscataway (2018)
26. Burattini, L., Bini, S., Burattini, R.: T-wave alternans quantification: which information from different methods? Comput. Cardiol. **37**, 1043–1046 (2010)

Chapter 8
Advanced Signal Processing Algorithms for Cardiorespiratory Monitoring in the Neonatal Intensive Care Unit

Riccardo Barbieri

Contents

8.1 Introduction

At the core of understanding advanced engineering and mathematical methods that can provide effective assisting technology in intensive care, the main role of the biomedical engineer is to approach the patients' monitored clinical state with the mathematical principles devised along these last centuries to characterize a dynamic physical system. As with many nonlinear dynamic systems, information indicating

R. Barbieri (✉)
Department of Electronics, Information and Bioengineering – Politecnico di Milano, Milan, Italy
e-mail: riccardo.barbieri@polimi.it; https://www.deib.polimi.it/eng/people/details/992020

© Springer Nature Switzerland AG 2021
D. Pani et al. (eds.), *Innovative Technologies and Signal Processing in Perinatal Medicine*, https://doi.org/10.1007/978-3-030-54403-4_8

the underlying parameters or state of a system can be well hidden within the observed signals (i.e., all the clinical variables monitored in the intensive care unit). In physiological systems, disentangling these underlying parameters can be complicated by the interaction of multiple subsystems.

In the neonatal intensive care unit (NICU), a critical step is to develop statistical techniques for the assessment of cardiovascular dynamics, cardiorespiratory interactions and respiratory instabilities in preterm infants, and more in general for the identification of any physiological state. The details of measurement modalities, time scales, and performance requirements may all differ significantly from one physiological problem to the next. However, it is likely that the insights and the stochastic-dynamic framework developed for a given problem help significantly in developing approaches for identification of other distinct physiological problems. Consequently, it is reasonable that a significant outcome of this research line will ultimately contribute a richer set of statistical tools and frameworks applicable to a wide range of physiological state identification problems.

8.2   Clinical Background

8.2.1   The Neonatal Intensive Care Unit

The term "neonatal" refers to the first 28 days of life. A neonatal intensive care unit (NICU), also known as an intensive care nursery (ICN), is an *intensive care unit* specializing in the care of ill or premature newborn infants. NICUs concentrate on treating very small, premature, or congenitally ill babies. Thanks to increasing technology, neonatology and NICUs have greatly increased the survival of very low birth-weight and extremely premature infants

One in 8 live births in the United States is preterm (<37 weeks post conception) and these high-risk births require specialized monitoring and treatment in NICUs. Apneic pauses causing transient hypoxia and associated bradycardia – often referred to as "cardiorespiratory events" – are common in preterm infants, with severity ranging from presumably benign periodic apnea with mild oxygen desaturations and cardiac decelerations to severe life-threatening apnea that requires mechanical ventilation. Although the severity and duration of cardiorespiratory events that require treatment are not established, prospective studies have linked intermittent hypoxia with the infant's current level of maturation, as well as a number of acute and long-term complications, including multiorgan dysfunction, retinopathy, developmental delays, and neuropsychiatric disorders. It is clear that apnea of prematurity is a major factor in prolonging hospitalization as well as raising concerns for subsequent risk of apparent life-threatening events and sudden infant death syndrome (SIDS) at home.

A major challenge for clinicians caring for preterm infants is that, while there is a multitude of physiological signals streaming across current NICU monitoring systems, there are very few techniques able to translate these signals into validated indices that define pathological states requiring treatment. Therefore, in the broadest sense, the critical role of the biomedical engineer is to develop and validate computational tools to embed in monitors that can provide real-time indices of cardiorespiratory stability and that can be linked to individualized time-sensitive interventions. Such monitoring system could be tailored to individual infants through a statistical framework that will quantify cardiorespiratory variability and infant vulnerability. The critical technical barriers are related to the development of parameter estimation techniques (discussed in methods). While preliminary results provide strong evidence that it is possible to monitor cardiorespiratory stability, the current understanding of the stochastic parameters is not sufficiently robust and accurate for a clinical setting. In order to refine these promising techniques, a rich and thorough long-term investigation is still required.

8.2.2 Common Diagnoses and Pathologies in the NICU

For sake of conciseness, here is a list of the main diagnoses and pathologies that have been observed in the NICU: anemia, apnea, bradycardia, bronchopulmonary dysplasia (BPD), hydrocephalus, intraventricular hemorrhage (IVH), jaundice, necrotizing enterocolitis (NEC), patent ductus arteriosus (PDA), periventricular leukomalacia (PVL), infant respiratory distress syndrome (RDS), retinopathy of prematurity (ROP), neonatal sepsis, and transient tachypnea of the newborn (TTN). We here focus on two main categories in connection with the main physiology that the chapter is aimed at outlining, i.e., cardiovascular control and control of breathing.

Life-threatening events are due to instabilities and malfunctioning of the cardiovascular and cardiorespiratory control system. It is believed that apneic events and poor respiratory function may also be contributing factors to sudden infant death syndrome. They also may have adverse consequences such as lengthening hospital stays, delaying development, or even irreparable damages that may affect the individual for his/her entire lifespan.

Relevant aspects related to this section can be found in [1–14].

8.3 Physiology of Cardiorespiratory Control in Infants

Life threatening events in the NICU are mainly due to instabilities and malfunctioning of the infant's cardiorespiratory control system. In preterm infants, decreases in heart rate (bradycardias) result in reduced cerebral blood velocity and delivery of oxygenated hemoglobin, as well as reduced clearance of metabolic byproducts. The total result of adverse cardiorespiratory events is hypoxic, i.e., there is an ischemic injury in tissue with high-metabolic demands. Intermittent hypoxia in preterm infants is associated with a range of complications including retinopathy, developmental delays, and neuropsychiatric disorders.

Heart rate is regulated by a feedback control system: blood pressure fluctuations are sensed by carotid sinus baroreceptors sending afferent impulses to the brainstem and the suprabulbar circuits. The consequent autonomic response regulates heart rate through vagal-sympathetic efferent nerves that affect the cardiac pacemaker. In pathological circumstances, the heart rate control system may be dysregulated, resulting in episodes of vagally mediated bradycardia.

The alternation and cyclicity of behavioral states in infancy is of particular interest since it provides an index of functioning and coordination of multiple neurological subsystems. In particular, sleep state plays a crucial role in autonomic control and maturation of the cardiorespiratory system. Within sleep epochs, it is possible to define distinct states characterized by different patterns of respiration, heart rate, electroencephalographic activity, eye and body movements, etc.

Since neonates spend the majority of their time sleeping, their sleep patterns are markedly different from the ones of older infants and adults. In early infancy, sleep states are classified as active sleep and quiet sleep, which can be seen as a first rudimentary version of the adult REM and non-REM sleep states, respectively.

In both sleep states, the respiratory rhythm is governed by neural circuits within the brainstem that signal the timing and depth of each breath. Continuous ventilation results from recurrent bursts of inspiratory neuronal activity that controls the diaphragm via discrete phrenic motor neuron activations. Infants with post-conceptional age of less than 36 weeks commonly have irregular breathing patterns with apneic events (periodic pauses in breathing). Preterm infant breathing patterns are highly non-stationary, with rapid changes in measures of breathing such as the time interval between breaths, called the inter-breath-interval (IBI), which is therefore an important measure for understanding irregularity of the breathing patterns.

Relevant aspects related to this section can be found in [15–35].

8.4   Methodology

A current line of work in the clinical intensive care unit (ICU) setting focuses on devising monitoring devices based on noninvasive recordings and able to characterize critical physiological mechanisms associated with cardiovascular control, the autonomic nervous system, and respiratory physiology, as well as to timely diagnose and possibly predict pathological states leading to disease.

When dealing with infants staying in the NICU, the main goal of the biomedical engineer is to provide mathematical methods and algorithms that use the instantaneously monitored physiological signals for predicting the occurrence of a life-threatening event (e.g., apnea, bradycardia, or neonatal sepsis). We here focus on specific methodologies that use the electrocardiogram (ECG), pulse plethysmogram and respiratory waveforms to extract meaningful information related to the physio-pathological state of the individual.

8.4.1 Methods for Cardiorespiratory Control Assessment: Heart Rate Variability in Adults

Despite the valuable research devoted to the study of cardiovascular mechanisms, there is still a need for specifically target standards and methods to assess the cardiorespiratory functions in the early stages of life.

Mathematical tools have been quite successful in quantifying important cardiovascular control mechanisms in adults. In particular, application of frequency domain methods to peak-to-peak series detected from the electrocardiogram (ECG) signal alone and together with cardiovascular covariates such as respiration and blood pressure have led to highly refined models and analysis tools, as well as successful efforts in defining specific standards, all now traditionally classified as "heart rate variability studies".

Heart rate is the number of R-wave events (heartbeats) per unit time. Heart rate variability is defined as the variation in either the R-R intervals (times between R-wave events) or heart rate with time. Heart rate variability is generated by autonomic control of the heart. Heart rate variability reflects healthy cardiovascular functions. Significant decreases in HRV has been reported in specific diseases like myocardial infarction, diabetic neuropathy, cardiac transplantation, hypertension, congestive heart failure, and fetal distress during labor.

As reported in 1996 by the seminal paper from the Task Force of the European Society of Cardiology and the North American Society of Pacing and Electrophysiology, from the *R-R* interval series we can compute simple time domain indices, like the mean NN interval, the mean heart rate, and the difference between the longest and shortest NN interval. More complex statistical time domain measures are also computed, which have been demonstrated to be useful in the study of HRV. Among these quantities are the standard deviation of all NN intervals (*SDNN*), the number of pairs of adjacent NN intervals differing by more than 50 ms divided by the total number of NN intervals (*pNN50*), and the square root of the mean of the sum of the squares of differences between adjacent NN intervals (*rMSSD*).

Traditional spectral HRV analysis is commonly carried out via periodogram analysis or autoregressive moving average (ARMA) estimation. Three main spectral components can be distinguished in a spectrum calculated from short-term recordings ranging from 2 to 20 minutes: very low frequencies (VLF: 0–0.04 Hz), low frequencies (LF: 0.04–0.15 Hz), high frequencies (HF: 0.15–0.4 Hz). The measurement of the spectral components is usually made in absolute values of power. LF and HF may also be measured in *normalized units* to emphasize the controlled and balanced behavior of the two branches of the autonomic nervous system. Vagal activity is the major contributor to the HF component, whereas disagreement exists with respect to the LF component. Some studies suggest that LF, when expressed in normalized units, is a quantitative marker of sympathetic modulations, while other studies have found that LF reflects both sympathetic and vagal activity. Consequently, the task force proposed the LF/HF ratio to mirror the synergic action of the sympatho-vagal balance, and thousands of published studies after their recommendation proved this index to provide the most indicative measure to quantify autonomic state.

When associating cardiovascular state with respiratory dynamics, the coherence function has been considered as a reasonable quantitative measure reflecting the strength of the linear interaction between HRV and respiration in adults. Coherence is traditionally calculated as the cross-spectral density between HRV and respiration normalized by the corresponding auto-spectral density functions. The coherence function takes values between zero, indicating absence of linear interactions, and one, indicating exclusive linear interactions. However, it has been pointed out that the estimation of coherence using cross-spectral density does not account for causality between the two variables considered, leading to the use of measures of Granger causality to assess directional influences.

Another possible source of information flow from HRV to respiratory control of ventilation could relate to neural reflexive baroreceptor influences on central neural respiratory activity. As a result, even with a non-negligible information flow in the opposite direction any interactions estimated using open loop paradigm can significantly differ from the actual interactions occurring in the closed loop. Such effect has been demonstrated in the interactions between heartbeat variations and systolic blood pressure values in adults.

To determine the significant interactions between HRV and respiration, a threshold level of coherence has been generally set either arbitrarily or based on statistical criteria derived from the sampling distribution. Any value of coherence above the threshold is considered as significant. Methods that are able to compute more appropriate coherence thresholds, theoretically or experimentally derived by the knowledge of estimator and signals under investigation, can avert discretionary use of a threshold not based on theoretical model or empirical approach.

To obtain frequency domain measures, both the R-R intervals and respiration are considered as output variables of a multivariate autoregressive model. The coefficients of the model are determined by solving the extended Yule-Walker equations and corresponding autospectra, coherence and gain are derived in the frequency domain from these coefficients. The statistical significance of the coherence for each infant is then determined by surrogate data analysis. Altogether, this approach provides a quantification of the linear relationship between R-R intervals and respiration, as well as its significance, defined along the entire range of frequencies. Consequently, specific indices for each of the LF and HF bands defined above can be derived from the spectral estimates.

Relevant aspects related to this section can be found in [36–50].

8.4.2 Methods for Cardiorespiratory Control Assessment: Heart Rate Variability in Infants

In infants, cardiorespiratory interactions are considered as an important indicator of the level of maturation of vagally mediated autonomic influence on the heart, although the precise relationship between HRV and respiration in preterm infants remains unknown.

The standard low frequency (LF: 0.04–0.15 Hz) and high frequency (HF: 0.15–0.4 Hz) ranges classified for adult HRV analysis do not apply to the analysis of HRV in infants. Attempts have been made to adapt new standards to newborn physiology. Different frequency ranges within 0.01 Hz–1.5 Hz are used for frequency domain analysis of HRV in preterm infants. Generally, any frequency above 0.2 Hz has been classified as HF in the case of infant HRV. However, a standard classification for infants has not yet been established.

As frequency domain indices have been established for adults by defining specific frequency ranges of interest according to breathing patterns, a critical point is to establish a similar classification of frequency ranges for infants. Starting from the observation that preterm infants have a predominant breathing frequency ~ 1 Hz, it makes sense to introduce a more refined characterization of the respiratory range, thereby classifying four different frequency ranges: the low frequency (LF: 0.01–0.15 Hz) and three high frequency ranges as (HF1: 0.15–0.45 Hz, HF2: 0.45–0.7 Hz, and HF3: 0.7–1.5 Hz). HF3 is generally at the frequency range corresponding to the eupneic respiratory rhythm of preterm infants.

The spectral coherence method assumes that the two signals interact in an open loop in which respiration has a unidirectional influence on the HRV whereas there may be information flow from HRV to respiration in preterm infants. For example, fluctuations in ventilation results in fluctuations in arterial pH and pCO_2, which in turn affect ventilatory drive via the central and peripheral chemoreceptors. The time delay and dynamics in this feedback system is modified by fluctuations in systemic and cerebral circulations, which are influenced by HRV. Correlated fluctuations in heart rate, arterial pressure cerebral circulation have been recorded in infants, and have been related to variability in breathing patterns of preterm infants. Therefore, information flow from HRV to respiration could be important in some infants making the interactions between HRV and respiration bi-directional and closed loop. In preterm infants, as cardiorespiratory interactions are weaker (or even absent) than adults, it is even more important to establish a solid statistical criterion to assess a reliable significance threshold for the coherence function.

The normal respiratory rate of infants is approximately 60 breaths per minute (1 Hz). However, most preterm infants have irregular breathing patterns with periodic breathing and pauses in breathing (apnea) that introduces frequencies lower than the normal range. As a result, the respiratory modulation of heart rate, if any, will be occurring at various ranges of frequencies, from ~1 Hz (normal breathing) and below. Hence, in preterm infants, the signature of RSA (i.e., the peak in the power spectrum of HRV at the normal breathing frequency of ~1 Hz) may not be observed in the HRV spectrum due to irregularity in breathing. Notably, a further complication in relying on traditional spectral analysis is that heart rate fluctuations may exist at the respiratory frequencies even in the absence of respiration.

Relevant aspects related to this section can be found in [51–63].

8.5 Exemplary Methodology: A Statistical Approach

The requirement of stationarity makes it more challenging to track changes in the temporal dynamics of heartbeat intervals. When dealing with non-stationarities, we may employ a time-varying moving window approach for a semi-continuous assessment of the time and frequency domain variables. These techniques try to approximate stationarity conditions by using specific filters and introducing concepts such as the forgetting factor.

Point process modeling provides a method to model the dynamic and stochastic processes of continuous systems defined by discrete observable events. A point-process is a stochastic process able to continuously characterize the intrinsic probabilistic structure of discrete events. It has been successfully applied to study a wide range of phenomena, analyzing data such as earthquake occurrences, traffic modeling, and neural spiking activity. In the case of respiration, the discrete events defining a point process framework are governed by neural circuits within the brainstem that signal the timing of each breath. Continuous ventilation results from recurrent bursts of inspiratory neuronal activity that control the diaphragm via discrete phrenic motor neuron activations. In the case of the heartbeat, the discrete events correspond to the electrical impulses from the heart's conduction system initiating ventricular contractions. The heartbeat generation mechanism can be modeled as a point process where ventricular contractions are discrete neuronal bursts governed by a complex control system involving the autonomic nervous system and the cardiac muscular system.

8.5.1 Basic Principles of Point Process Modeling

A *temporal point process* is a stochastic time-series of events that occurs in continuous time. A point-process can be represented by the timing of the events, by the waiting times between events, using a counting process, as a set of 1s and 0s, very similar to binary (if time is discretized enough to ensure that in each window only one event has the possibility of occurring, that is to say one time bin can only contain one event).

In the binary representation, the point process can only take on *two values at each point in time*, indicating whether or not an event has actually occurred. In this way, what carries the actual event generation information is the *occurrence of an event*, as well as the time between successive events. Using this approach, it is possible to retrieve the flow of information from the autonomic pathways through an observation period.

The *rate function* λ of a Poisson process (the fixed mean rate of the event occurrence) defines events occurring in non-overlapping intervals that are independent. In this case, the inter-event-interval (IEI) probability density is the exponential probability density. In inhomogeneous Poisson processes the rate function is time-varying ($\lambda = \lambda(t)$). Also in the inhomogeneous case, if events occurring in non-overlapping intervals are independent, the inter-event-interval probability density is the exponential probability density.

Generalizing the simple Poisson model The Poisson process is limited in that it is memory-less. It does not account for any event history when calculating the current occurrence probability. Biological events exhibit a fundamental (biophysical) history dependence by way of their relative and absolute refractory periods.

There are two ways to generalize a simple Poisson process in order to construct more accurate models underlying the event generation:

- Generalize the rate function to get an inhomogeneous poisson process defined by a conditional intensity function.
- Generalize the inter-event-interval distribution to obtain the inhomogeneous poisson process as a renewal process

The conditional intensity function To address history dependence, a conditional intensity function is used to represent the probability of an event occurrence, conditioned on its own history. The conditional intensity function expresses the instantaneous occurrence probability and implicitly defines a complete probability model for the point process. It defines a probability per unit time. If this unit time is taken small enough to ensure that only one event could occur in that time window, then our conditional intensity function completely specifies the probability that a given event occurs at a specific time.

Any probability density satisfying $f(t) > 0$ for $t > 0$ can be considered as a renewal probability density. Probability models used as renewal processes are so common that exhaustive related information can be found on Wikipedia. They include:

The Exponential Distribution Probability distribution that describes the time between events in a *Poisson process*, i.e., a process in which events occur continuously and independently at a constant average rate. It is a particular case of the gamma distribution. It is the continuous analog of the geometric distribution, and it has the key property of being memory-less.

The Gamma Distribution The gamma distribution is a two-parameter family of continuous probability distributions. The common exponential distribution and chi-squared distribution are special cases of the gamma distribution. There are three main parametric representations of the distribution:

With a shape parameter k and a scale parameter θ

With a shape parameter $\alpha = k$ and an inverse scale parameter $\beta = 1/\theta$, called a *rate parameter*

With a shape parameter k and a mean parameter $\mu = k/\beta$

The Log-Normal Distribution A log-normal (or lognormal) distribution is a continuous probability distribution of a random variable whose logarithm is normally distributed. Thus, if the random variable X is log-normally distributed, then $Y = \ln(X)$ has a normal distribution. Likewise, if Y has a normal distribution, then $X = \exp(Y)$ has a log-normal distribution. A random variable which is log-normally distributed takes only positive real values.

The Inverse Gaussian Distribution The inverse Gaussian distribution is a two-parameter family of continuous probability distributions with support on $(0, \infty)$. The first parameter is the mean of the distribution, the second parameter is a shape parameter $K > 0$. As K tends to infinity, the inverse Gaussian distribution becomes more like a normal (Gaussian) distribution. The inverse Gaussian has several properties analogous to a Gaussian distribution. The name can be misleading: it is an "inverse" only in that, while the Gaussian describes a Brownian motion's level at a fixed time, the inverse Gaussian describes the distribution of the time a Brownian motion with positive drift takes to reach a fixed positive level. For this reason, the inverse Gaussian distribution is the link between deterministic and stochastic models of neural spiking activity because it can be derived from an *integrate and fire model* defined as a random walk with drift where the additive noise is a Wiener process (Brownian motion).

The Local Maximum Likelihood Approach To calculate the local maximum likelihood estimate of all the parameters (summarized within a general vector θ), we define the local joint probability density associated with $f(u_{t-1:t}| \theta_t)$ within the length of the local likelihood observation interval. If we observe n_t peaks within this interval as $u_1 < u_2 <, \ldots\ldots, < u_{n_t} \leq t$ and if the paramenter in θ are time varying, then at time t, we estimate the maximum likelihood estimate of $\hat{\theta}_t$ to be the estimate of θ in the interval l. Considering the right censoring, the local log likelihood is obtained as

$$\log f\left(u_{t-1:t} \mid \theta_t\right) = \sum_{i=2}^{n_t} w\left(t-u_i\right)\log f\left(u_i - u_{i-1} \mid H_{u_{i-1}}, \theta_t\right)$$
$$+ w\left(t-u_{n_t}\right)\log \int_{t-u_{n_t}}^{\infty} f\left(\vartheta \mid H_{u_{n_t}}, \theta_t\right)d\vartheta$$

where $w(t)$ is a weighting function to account for faster updates to local likelihood estimation and we selected as $w(t) = e^{-\alpha(t-u)}$ with α as the weighting time constant that assigns the influence of a previous observation on the local likelihood at time

t. Since θ can be estimated in continuous time, we can obtain the instantaneous estimate of μ, the mean, using the autoregressive representation. Similarly, the local likelihood estimate can also provide the instantaneous estimate of the second moment of the distribution.

Model Goodness-of-Fit The IBI probability model along with the local maximum likelihood method provides an approach for estimating the instantaneous mean and instantaneous variance of the IBI. These measures provide information about the changes in the characteristics of the distribution, possibly due to the irregularity of breathing. However, it is also essential to evaluate how well the model represents the IBI. To obtain a goodness-of-fit measure we compute the time-rescaled IBI defined as

$$\tau_k = \int_{u_{k-1}}^{u_k} \lambda\left(t \mid H_t, \hat{\theta}_t\right) dt$$

where the u_k represent the breathing events observed in (0,T) and $\lambda\left(t \mid H_t, \hat{\theta}_t\right)$ is the conditional intensity function defined as

$$\lambda\left(t \mid H_t, \hat{\theta}_t\right) = f\left(t \mid H_t, \hat{\theta}_t, \hat{\sigma}_t\right)\left[1 - \int_{u_{n_t}}^{t} f\left(\vartheta \mid H_\vartheta, \hat{\theta}_\vartheta, \hat{\sigma}_\vartheta\right) d\vartheta\right]^{-1}$$

The conditional intensity is the history-dependent rate function for a point process that generalizes the rate function for a Poisson process. According to the *Time Rescaling Theorem*, any set of observations from a point process that has a conditional intensity function can be transformed into a sequence of independent exponential random variables with a rate of 1.

Therefore, the τ_k values are independent, exponential random variables with a unit rate. With a transformation $z_k = 1 - \exp\left(-\tau_k\right)$, the z_k values become independent, uniform random variables on the interval (0,1]. Thus, we can employ a KS test to assess the agreement between the transformed z_k values and a uniform probability density. If there is close agreement between the point process model and the IBI data series, then the transformed z_k values plotted against the uniform density will have close agreement if the plot is closer to the 45 degrees diagonal (KS plot). The KS distance measures the largest distance between the cumulative distribution function of the IBI transformed in the interval (0,1] and the cumulative distribution function of a uniform distribution on (0,1]. The smaller the KS distance, the better the model in terms of goodness-of-fit.

Relevant aspects related to this section can be found in [64–75].

8.5.2   A Point Process Model of Cardiovascular Dynamics

More recently the utility of point process theory has been validated as a powerful tool to estimate heartbeat and respiratory dynamics, including instantaneous measures of variability and stability, and in short recordings under nonstationary conditions. In contrast, the commonly used standard methods are primarily applicable for stationary data or provide only approximate estimates of the dynamic signatures that are not corroborated by goodness-of-fit methods. The few methods available for time-frequency analysis for nonstationary data (e.g., Hilbert-Huang and Wavelet transforms) need to be applied to short batches of data, making them less suitable for tracking dynamics in real time. Finally, the point process framework allows for inclusion of any covariate at any sampling rate, and this property can be used to generate instantaneous indices of cardiovascular and respiratory variability.

The *R-R* interval is the time interval between successive heart contractions, which are depicted as the *R* peaks of the QRS complex on the ECG. If we consider a data collection interval $[T_a, T_b]$, the *R-R* peaks within this window are given by: $T_a \leq u_1 < u_2 < \ldots < u_i < \ldots < u_k \leq T_b$, where each u_i is the time of the *ith R* peak. Thus, the corresponding R-R time interval at time k is given by the set $H_k = \{w_k, w_{k-1}, \ldots, w_i, \ldots, w_{k-p+1}\}$, where $w_k = u_k - u_{k-1}$ and $p \leq i \leq k$.

Because cardiac contraction is a serial procedure where the occurrence of a heartbeat may be influenced by previous contractions (e.g., action potential and muscle contraction refractory times), we can attempt to model a heartbeat at time k with a *p*-order autoregressive process:

$$\mu(H_k, \theta) = \theta_o + \sum_{j=1}^{p} \theta_j w_{k-j+1} + \epsilon_k$$

where $\theta = \{\theta_o, \ldots, \theta_j, \ldots, \theta_k\}$ is the estimation vector of optimized model parameters and ϵ_k. Is usually defined as Gaussian white noise.

There are two facts that suggest a more refined statistical approach. First of all, heartbeat intervals are the times between two events (the *R*-wave events). These events correspond to the electrical impulses from the heart's conduction system, which initiate ventricular contractions. Therefore, *R*-wave events are a sequence of discrete occurrences in continuous time and hence, they form a point process. Secondly, the autonomic nervous system is the principal dynamic system that modulates the dynamics of the heartbeat intervals. These facts taken together suggest that heartbeat interval measurements can be analyzed meaningfully using a more complex probabilistic model of a dynamical system observed through a point process. In this model, the observation equation summarizes the stochastic properties of the observed heartbeat point process, whereas the essential features of the parasympathetic and sympathetic activity will be concisely summarized in a history-dependent, time-varying structure (that can be for example the regression formulated above).

Since the probability density description that arises from the model characterizes the stochastic properties of the *R-R* interval, we use it to formulate precise definitions of heart rate and heart rate variability. Instantaneous heart rate is often defined as the reciprocal of the *R-R* intervals. Hence, for any particular point in time, we define the associated instantaneous heart rate as the inverse of the waiting time until the next *R*-wave event up to a constant that converts the *R-R* interval measurements recorded in milliseconds into heart rate measurements reported in beats per minute (bpm). We can derive the probability density of the instantaneous heart rate from the

R-R interval probability density by using the change-of-variable formula from elementary probability theory. This density then defines the stochastic properties of heart rate.

To use this model in real data analysis, heart rate should be a representative value from the instantaneous heart rate probability density. Therefore, we define heart rate as the mean of this density, and heart rate variability as the standard deviation. Instantaneous assessment of these indices can be performed using either a local likelihood algorithm, or an adaptive point process filter.

Our instantaneous measurements can be computed simultaneously from a single statistical framework, they are computed in continuous time, and they can be extracted at any time resolution. Previous methods compute similar estimates either on a beat-to-beat basis, or in continuous time by preprocessing and filtering of the original *R-R* interval series not justified by a physiological model of heartbeat generation as in our case. Our previous studies also suggest that summaries comparable to *SDNN* and *LF/HF* analyses can be performed with heart rate series derived from our history-dependent inverse Gaussian (HDIG) model. These results coupled with the goodness-of-fit analyses, which demonstrated that point process models offer an accurate description of the stochastic structure in the heartbeat interval series, suggest that static and dynamic measures derived from point process methods may be a more accurate description of these quantities.

The Point Process Model We assume that given any *R*-wave event u_k,, the waiting time until the next *R*-wave event, or equivalently, the length of the next *R-R* interval, obeys a HDIG probability density $f\left(t \mid H_{u_k}, \theta\right)$, where t is any time satisfying $t > u_k$, H_{u_k} is the history of the *R-R* intervals up to u_k, and θ is a vector of model parameters. The model is defined as

$$f\left(t \mid, H_{u_k} \mid, \theta\right) = \left[\frac{\theta_{p+1}}{2\pi\left(t - u_k\right)^3}\right]^{\frac{1}{2}} \exp\left\{-\frac{1}{2}\frac{\theta_{p+1}\left[t - u_k - \mu\left(H_{u_k}, \theta\right)\right]^2}{\mu\left(H_{u_k}, \theta\right)^2\left(t - u_k\right)}\right\}$$

where $H_k = \{u_k, w_k, w_{k-1}, \ldots, w_i, \ldots, w_{k-p+1}\}$, $w_k = u_k - u_{k-1}$ is the *Kth R-R* interval,

$$\mu\left(H_k, \theta\right) = \theta_o + \sum_{j=1}^{p} \theta_j w_{k-j+1}$$

is the mean, $\theta_{p+1} > 0$ is the scale parameter, and $\theta = \{\theta_o, \ldots, \theta_j, \ldots, \theta_k\}$. This model represents the dependence of the *R-R* interval length on the recent history of parasympathetic and sympathetic inputs to the sinoatrial node by modeling the mean as a linear function of the last p *R-R* intervals. If we assume that the *R-R* intervals are independent (i.e., $p = 0$), then $\mu\left(H_{u_k}, \theta\right) = \theta_0$, $f\left(t \mid H_{u_k}, \theta\right) = f\left(t \mid u_k, \theta_o, \theta_1\right)$, and the equation simplifies to a renewal inverse Gaussian (RIG) model. The mean and standard deviation of the *R-R* interval probability model are respectively,

$$\mu_{RR} = \mu\left(H_{u_k},\theta\right)$$

$$\sigma_{RR} = \left[\mu\left(H_{u_k},\theta\right)^3 \theta_{p+1}^{-1}\right]^{\frac{1}{2}}$$

Because our probability density characterizes the stochastic properties of the R-R intervals, we use it to formulate precise definitions of heart rate and heart rate variability. As mentioned in the task force report, heart rate is often defined as the reciprocal of the R-R intervals. Hence, for any $t > u_k$, $t - u_k$, is the waiting time until the next R-wave event, and we can define $r = c(t - u_k)^4$, as the heart rate random variable, where $c = 6 * 10^4$, msec/min is the constant that converts the R-R interval measurements in milliseconds into heart rate measurements in beats per minute (bpm). Therefore, because r is a one-to-one transformation of $t - u_k$, we use the standard change-of-variables formula from elementary probability theory (Ross 1997) and derive from the R-R interval probability density in, $f\left(r\,|\,H_{u_k},\theta\right)$, the heart rate probability density defined as

$$f\left(r\,|,H_{u_k}\,|,\theta\right) = \left|\frac{dt}{dr}\right| f\left(t\,|,H_{u_k}\,|,\theta\right) = \left[\frac{\theta_{p+1}^*}{2\pi r}\right]^{\frac{1}{2}} \exp\left\{-\frac{1}{2}\frac{\theta_{p+1}^*\left[1-\mu^*\left(H_{u_k},\theta\right)r\right]^2}{\mu^*\left(H_{u_k},\theta\right)^2 r}\right\},$$

where $\mu^*\left(H_{u_k},\theta\right) = c^{-1}\mu\left(H_{u_k},\theta\right)$ and $\theta_{p+1}^* = c^{-1}\theta_{p+1}$. The mean and standard deviation of the heart rate probability density are respectively:

$$\mu_{HR} = \mu^*\left(H_{u_k},\theta\right)^{-1} + \theta_{p+1}^{*-1}$$

$$\sigma_{HR} = \left[\frac{2\mu^*\left(H_{u_k},\theta\right) + \theta_{p+1}^*}{\mu^*\left(H_{u_k},\theta\right)\cdot\theta_{p+1}^{*2}}\right]^{\frac{1}{2}}$$

Our R-R interval probability model provides an approach for estimating instantaneous mean R-R interval, heart rate, R-R interval standard deviation, and heart rate standard deviation from a time-series of R-R intervals. Therefore, our framework provides new ways for estimating heart rate and heart rate variability and for assessing model goodness-of-fit by considering formally the point process structure in the data.

Embedding the Autoregressive Model on the R-R Intervals Computational procedures based on a comparison of the prediction power of linear and nonlinear models of the Volterra-Wiener form have been applied to continuous time series to measure deterministic and chaotic dynamics of heartbeats. Including nonlinear terms of past *R-R* intervals usually improves our model fits. The mean of the probability function in this case is redefined as

$$\mu\left(H_{u_k},\theta\right)=\theta_0+\sum_{j=1}^{p}\theta_j w_{k-j+1}+\sum_{i=1}^{q}\sum_{j=1}^{q}\phi_{ij}w_{k-i+1}w_{k-j+1}>0$$

This formulation can also be interpreted as a discrete Volterra-Wiener-Koremberg series of degree of nonlinearity $d = 2$ and memory $h = $ max (p, q). Both the local maximum likelihood and the adaptive filter algorithms will be applied for model fitting. The importance of the nonlinear parameters in comparison with the linear terms, together with goodness-of-fit measurements, will give a measure of nonlinearity of the point process generating the heartbeats. The linear and nonlinear indices of HRV will be defined as a function of the parameters $\theta = \{\theta_o, ..., \theta_j, ..., \theta_k\}$ and $\phi = \{\phi_{11}...\phi_{qq}\}$ respectively. Application of this paradigm will allow us to investigate if the degree of nonlinearity is dependent on the physiological state of the cardiovascular system.

Including other Cardiovascular Variables as Covariates We can also describe a more complete model if we include dependence not only from past beat intervals, but also on external covariates involved in cardiovascular control. The mean of the probability function becomes

$$\mu\left(H_{u_k},\theta\right)=\theta_0+\sum_{j=1}^{p}\theta_j w_{k-j+1}+\sum_{j=1}^{q}\gamma_j \text{COV}_{k-j+1}+....>0$$

where the number of covariates may be more than one. Since they are considered together with autoregressions on the *R-R* intervals, their values are sampled in correspondence to the beat series. A similar formulation with the autoregression on the mean *R-R* intervals will allow for consideration of covariate time series.

Point process modeling techniques have been indeed used to investigate infant physiology. Results have shown that the lognormal probability distribution is sufficient in modeling the instabilities in underdeveloped infant cardiovascular physiology. Therefore, we assume that a collection of *R-R* intervals is a log-normally distributed random variable. At any given peak, u_k, we model the waiting time until the next heartbeat, u_{k+1}, with a lognormal probability density:

$$f_{k+1}\left(t\mid H_k,\theta\right)=\left[\frac{1}{2\pi\sigma^2\left(t-u_k\right)^2}\right]^{\frac{1}{2}}\exp\left\{-\frac{1}{2}\frac{\left(\ln\left(t-u_k\right)-\mu\left(H_k,\theta\right)\right)^2}{\sigma^2}\right\}$$

Where, for a given time $t < u_{k+1}$, H_k is the set of all *R-R* intervals prior to u_k, and $\mu(H_k, \theta)$ and σ represent the logarithmic form of the mean and standard deviation of the sample distribution. All parameters (and consequent indices) can be then estimated over time as shown in Sect. 8.5.4.

8.5.3   A Point Process Model of Respiratory Dynamics

A basic assumption of a statistical model for breathing is that the peak of inspiration, marked by the peak of inhalation recorded non-invasively, is a discrete event
that marks the timing of neuronal inspiratory bursts. An additional assumption
needed to use a point process paradigm is that IBI dynamics are governed by continuous processes under the regulation of multiple feedbacks and loops acting upon
the respiratory oscillator.

As a starting point, we hypothesize that the IBI of the infant follows a power law
distribution, and the characterizing parameters of the distribution are found to be
sensitive to age (maturation) [4]. We considered in an observation interval $(0, T]$,
successive peaks of the respiratory signal, $0 < u_1 < u_2 < ,\ \ldots\ldots\ldots, < u_k < ,$
$\ldots\ldots\ldots, < u_K \leq T$. Then, we assume that at any given peak u_k, the waiting time until
the next peak obeys a history-dependent lognormal probability density $f(t| H_k, \theta)$ as

$$f_{k+1}\left(t \mid H_k,\theta\right) = \left[\frac{1}{2\pi\sigma^2\left(t-u_k\right)^2}\right]^{\frac{1}{2}} \exp\left\{-\frac{1}{2}\frac{\left(\ln\left(t-u_k\right)-\mu\left(H_k,\theta\right)\right)^2}{\sigma^2}\right\}$$

where t is any time, $t > u_k$, H_k is the history of IBI up to u_k represented as $H_k = \{u_k,$
$w_k, w_{k-1}, \ldots, w_{k-p+1}\}$ with $w_k = u_k - u_{k-1}$ is the kth IBI and θ is a vector of model
parameters. The instantaneous mean is modeled as a p-order autoregressive process
as $\mu\left(H_k,\theta\right) = \theta_o + \sum_{j=1}^{p} \theta_j w_{k-j+1}$. The probability density in the equation defines the
IBI distribution with μ and σ as the characterizing parameters. At each instant of
time t, to estimate θ and σ, we can employ the local maximum-likelihood approach
defined in Sect. 8.5.4

8.5.4   A Statistical Model of Cardio-Respiratory Dynamics

Bivariate Autoregressive Analysis of HRV and Respiration A bivariate autoregressive model is employed to study the interaction between RR and respiration
(RP). The model is defined as

$$X\left(n\right) = -\sum_{k=1}^{M} A\left(k\right).X\left(n-k\right) + w\left(n\right)$$

$$n = 1, 2, 3 \ldots\ldots\ldots, N$$

where M is the order and is set at 32, N is the total number of data points,

$$X(n) = \begin{bmatrix} RR(n) & RP(n) \end{bmatrix}, A(k) = \begin{bmatrix} a_{11}(k) & a_{12}(k) \\ a_{21}(k) & a_{22}(k) \end{bmatrix}$$

and

$$w(n) = \begin{bmatrix} w_{RR}(n) & w_{RP}(n) \end{bmatrix}.$$

where $w(n)$ represents the white noise and $a_{ij}(k)$ represents the autoregressive coefficients. Clearly, this formulation provides the most simple statistical structure where the uncertainty of the outcome is modeled as a Gaussian noise and the first two moments of the distribution univocally define the distribution function of the model uncertainty.

Of note, we can use a recursive algorithm to determine the coefficients of the autoregressive model; spectral components are determined from these coefficients [15].

In the frequency domain,n the model is represented as:

$$\begin{bmatrix} RR(f) \\ RP(f) \end{bmatrix} = \begin{bmatrix} A_{11}(f) & A_{12}(f) \\ A_{21}(f) & A_{22}(f) \end{bmatrix} \begin{bmatrix} RR(f) \\ RP(f) \end{bmatrix} + \begin{bmatrix} w_{RR}(f) \\ w_{RP}(f) \end{bmatrix}$$

where $A_{ij}(f) = \sum_{M}^{k=1} a_{ij}(k) e^{-l2\pi fk}$ with $i, j = 1, 2$ and $l = \sqrt{-1}$, a complex quantity
The equation can be reformulated as:

$$\begin{bmatrix} RR(f) \\ RP(f) \end{bmatrix} = \begin{bmatrix} h_{11}(f) & h_{12}(f) \\ h_{21}(f) & h_{22}(f) \end{bmatrix} \begin{bmatrix} w_{RR}(f) \\ w_{RP}(f) \end{bmatrix}$$

where

$$h_{11}(f) = \frac{1 - A_{22}(f)}{\left(1 - A_{11}(f)\right)\left(1 - A_{22}(f)\right) - A_{21}(f) A_{12}(f)}$$

$$h_{12}(f) = \frac{A_{12}(f)}{\left(1 - A_{11}(f)\right)\left(1 - A_{22}(f)\right) - A_{21}(f) A_{12}(f)}$$

$$h_{21}(f) = \frac{A_{21}(f)}{\left(1 - A_{11}(f)\right)\left(1 - A_{22}(f)\right) - A_{21}(f) A_{12}(f)}$$

$$h_{22}(f) = \frac{1 - A_{11}(f)}{\left(1 - A_{11}(f)\right)\left(1 - A_{22}(f)\right) - A_{21}(f) A_{12}(f)}$$

The coherence γ^2 at a specific frequency f is evaluated using the classical definition as

$$\gamma^2(f) = \frac{|P_{CROSS}(f)|^2}{P_{RR}(f)P_{RP}(f)}$$

where $P_{RR}(f)$ and $P_{RP}(f)$ are the auto-spectral density functions of RR and RP respectively. $P_{CROSS}(f)$ is the cross-spectral density between RR and RP. The auto-and cross-spectral density functions are evaluated as

$$\begin{bmatrix} P_{RR}(f) & P_{CROSS}(f) \\ P_{CROSS}(f) & P_{RP}(f) \end{bmatrix} = \begin{bmatrix} |h_{11}|^2\,\sigma_{RR}^2 + |h_{12}|^2\,\sigma_{RP}^2 & h_{11}^*h_{21}\sigma_{RR}^2 + h_{12}^*h_{22}\sigma_{RP}^2 \\ h_{21}^*h_{21}\sigma_{RR}^2 + h_{22}^*h_{12}\sigma_{RP}^2 & |h_{21}|^2\,\sigma_{RR}^2 + |h_{22}|^2\,\sigma_{RP}^2 \end{bmatrix}$$

The causal coherence is calculated using the same Eq. (8.3) with the corresponding loop set to zero, thus for the respiration to RR causal coherence (RP $\rightarrow$ RR), we set $h_{21} = 0$ and for RR to respiration causal coherence (RR $\rightarrow$ RP) $h_{12} = 0$. Similarly, corresponding gains are calculated as

$$\text{Gain}(RP \rightarrow RR) = \left|\frac{h_{12}(f)}{h_{22}(f)}\right| = \left|\frac{A_{12}(f)}{1 - A_{11}(f)}\right|$$

$$\text{Gain}(RR \rightarrow RP) = \left|\frac{h_{21}(f)}{h_{11}(f)}\right| = \left|\frac{A_{21}(f)}{1 - A_{22}(f)}\right|$$

It has to be noted that these estimated causal gains are relevant only if the corresponding causal coherence values are significant. Hence detecting the significant coherence values between RR and respiration is an important step in establishing the presence of interactions in preterm infants.

As both coherence and gain are estimated along the entire frequency range up to the Nyquist frequency, we can further compute the maximum coherence and the corresponding gain in each of the defined frequency band (LF, HF1, HF2, or HF3).

Relevant aspects related to this section can be found in [76–88].

References

1. Griffin, M.P., Moorman, J.R.: Toward the early diagnosis of neonatal Sepsis and Sepsis-like illness using novel heart rate analysis. Pediatrics. **107**(1), 97–104 (2001)
2. Pichler, G., Urlesberger, B., Müller, W.: Impact of bradycardia on cerebral oxygenation and cerebral blood volume during apnoea in preterm infants. Physiol. Meas. **24**(3), 671–680 (2003)
3. Janvier, A., et al.: Apnea is associated with neurodevelopmental impairment in very low birth weight infants. J. Perinatol. **24**, 763–768 (2004)

4. National Center for Health Statistics (U.S.): Public Use Data Tapes from the National Center for Health Statistics Set: 1992–2002. U.S. Department of Health and Human Services, Public Health Service, Centers for Disease Control, National Center for Health Statistics, Hyattsville (2005)

5. Abu-Shaweesh, J.M., Martin, R.J.: Neonatal apnea: what's new? Pediatr. Pulmonol. **43**(10), 937–944 (2008)

6. Poets, C.E.: Interventions for apnoea of prematurity: a personal view. Acta Paediatr. **99**(2), 172–177 (2010)

7. Poets, C.E.: Apnea of prematurity: what can observational studies tell us about pathophysiology? Sleep Med. **11**(7), 701–707 (2010)

8. Di Fiore, J.M., Bloom, J.N., Orge, F., Schutt, A., Schluchter, M., Cheruvu, V.K., Walsh, M., Finer, N., Martin, R.J.: A higher incidence of intermittent hypoxemic episodes is associated with severe retinopathy of prematurity. J. Pediatr. **157**(1), 69–73 (2010)

9. Zhao, J., Gonzalez, F., Mu, D.: Apnea of prematurity: from cause to treatment. Eur J Pediatr. **170**(9), 1097–1105 (2011)

10. Mathew, O.P.: Apnea of prematurity: pathogenesis and management strategies. J. Perinatol. **31**(5), 302–310 (2011)

11. Martin, R.J., Wang, K., Köroğlu, Ö., Di Fiore, J., Kc, P.: Intermittent hypoxic episodes in preterm infants: do they matter? Neonatology. **100**(3), 303–310 (2011)

12. Poets, F.: Interventions for apnea of prematurity: a personal view. Acta Paediatr. **99**(2), 172–177 (2010)

13. Picone, S., Bedetta, M., Paolillo, P.: Caffeine citrate: when and for how long. A literature review. J. Matern. Fetal Neonatal Med. **25**, 11–14 (2012)

14. Aylward, G.: Neurodevelopmental outcomes of infants born prematurely. J. Dev. Behav. Pediatr. **35**(6), 394–407 (2014)

15. Prechtl, H.F.R.: The behavioural states of the newborn infant (a review). Brain Res. **76**(2), 185–212 (1974). Eckberg DL, Kifle YT, Roberts VL. Phase relationship between normal human respiration and baroreflex responsiveness. J Physiol London 1980;304:489–502

16. Vyas, H., et al.: Relationship between apnoea and bradycardia in preterm infants. Acta Paediatrica Scand. **70**, 785–790 (1981)

17. Waggener, T.B., Frantz III, I.D., Stark, A.R., Kronauer, R.E.: Oscillatory breathing patterns leading to apneic spells in infants. J. Appl. Physiol. **52**(5), 1288–1295 (1982)

18. Waggener, T.B., Frantz, I.D., Stark, A.R., Kronauer, R.E.: Oscillatory breathing patterns leading to apneic spells in infants. J. Appl. Physiol. **52**, 1288–1295 (1982)

19. Perlman, J.M., Volpe, J.J.: Episodes of apnea and bradycardia in the preterm newborn: impact on cerebral circulation. Pediatrics. **76**, 333–338 (1985)

20. Carley, D.W., Shannon, D.C.: Relative stability of human respiration during progressive hypoxia. J. Appl. Physiol. **65**, 1389–1399 (1988)

21. Berntson, G.G., Cacioppo, J.T., Quigley, K.S.: Respiratory sinus arrhythmia—autonomic origins, physiological-mechanisms, and psycho physiological implications. Psychophysiology. **30**, 183–196 (1993)

22. Paydarfar, D., Buerkel, D.M.: Dysrhythmias of the respiratory oscillator. Chaos. **5**, 18–29 (1995)

23. Paydarfar, D., Buerkel, D.M.: Collapse of homeostasis during sleep. In: Schwartz, W.J. (ed.) Sleep Science: Integrating Basic Research and Clinical Practice, pp. 60–85. Karger, Basel (1997)

24. Frey, U., Silverman, M., Baraba'si, A.L., Suki, B.: Irregularities and power law distributions in the breathing pattern in preterm and term infants. J. Appl. Physiol. **85**, 789–797 (1998)

25. Khoo, M.C.K.: Determinants of ventilatory instability and variability. Respir. Physiol. **122**, 167–182 (2000)

26. Mortola, J.P.: Respiratory Physiology of Newborn Mammals: a Comparative Perspective. The Johns Hopkins Press, Baltimore (2001)

27. Del Negro, C.A., Wilson, C.G., Butera, R.J., Rigatto, H., Smith, J.C.: Periodicity, mixed-mode oscillations, and quasiperiodicity in a rhythm-generating neural network. Biophys. J. **82**, 206–214 (2002)

28. Lehtonen, L., Martin, R.J.: Ontogeny of sleep and awake states in relation to breathing in preterm infants. Semin. Neonatol. **9**, 229–238 (2004)
29. Feldman, L., Del Negro, C.A.: Looking for inspiration: new perspectives on respiratory rhythm. Nat. Rev. Neurosci. **7**(3), 232–242 (2006)
30. Bruce, E.N.: Temporal variations in the pattern of breathing. J. Appl. Physiol. **80**, 1079–1087 (1996)
31. Darnall, R.A., Ariagno, R., Kinney, H.C.: The late preterm infant and the control of breathing, sleep, and brainstem development: a review. Clin. Perinatol. **33**, 883–914 (2006)
32. Patural, H., Pichot, V., Jaziri, F., Teyssier, G., Gaspoz, J.M., Roche, F., et al.: Autonomic cardiac control of very preterm newborns: a prolonged dysfunction. Early Hum. Dev. **84**, 681–687 (2008). Tarullo, A.R., Balsam, P.D., Fifer, W.P.: Sleep and infant learning. Infant Child Dev. **20**(1), 35–46 (2011)
33. Williamson, J.R., et al.: Forecasting respiratory collapse: theory and practice for averting life-threatening infant apneas. Respir. Physiol. Neurobiol. **189**, 223–231 (2013)
34. Poets, C.F., et al.: Association between intermittent hypoxemia or bradycardia and late death or disability in extremely preterm infants. JAMA. **314**, 595–603 (2015)
35. Di Fiore, J.M., et al.: Cardiorespiratory events in preterm infants: interventions and consequences. J. Perinatol. **36**, 251–258 (2016)
36. Akselrod, S., Gordon, D., Ubel, F.A., Shannon, D.C., Barger, A.C., Cohen, R.J.: Power spectrum analysis of heart-rate-fluctuation—a quantitative probe of beat-to-beat cardiovascular control. Science. **213**, 220–222 (1981)
37. Priestley, M.B.: Spectral Analysis and Time Series. Academic Press, New York (1983)
38. Kay, S.M.: Modern Spectral Estimation. Prentice Hall, Englewood Cliffs (1988)
39. Malik, M., Farrell, T., Cripps, T., Camm, A.J.: Heart-rate-variability in relation to prognosis after myocardial-infraction—selection of optimal processing techniques. Eur. Heart J. **10**, 1060–1074 (1989)
40. Saul, J.P., Berger, R.D., Albrecht, P., Stein, S.P., Chen, M.H., Cohen, R.J.: Transfer-function analysis of the circulation—unique insights into cardiovascular regulation. Am. J. Phys. **261**, H1231–H1245 (1991)
41. Theiler, J., Eubank, S., Longtin, A., Galdrikian, B., Farmer, J.D.: Testing for nonlinearity in time-series—the method of surrogate data. Physica D. **58**, 77–94 (1992)
42. Camm, A.J., Malik, M., Bigger, J., Breithardt, G., Cerutti, S., Cohen, R., Coumel, P., Fallen, E., Kennedy, H., Kleiger, R., et al.: Heart rate variability: standards of measurement, physiological interpretation and clinical use. Task Force of the European Society of Cardiology and the North American Society of Pacing and Electrophysiology. Circulation. **93**(5), 1043–1065 (1996)
43. Berntson, G.G., Bigger, J.T., Eckberg, D.L., Grossman, P., Kaufmann, P.G., Malik, M., et al.: Heart rate variability: origins, methods, and interpretive caveats. Psychophysiology. **34**, 623–648 (1997)
44. Chon, K.H., Mukkamala, R., Toska, K., Mullen, T.J., Armoundas, A.A., Cohen, R.J.: Linear and nonlinear system identification of autonomic heart-rate modulation. IEEE Eng. Med. Biol. Mag. **16**(5), 96–105 (1997)
45. Schafer, C., Rosenblum, M.G., Kurths, J., Abel, H.H.: Heartbeat synchronized with ventilation. Nature. **392**, 239–240 (1998)
46. Mancia, G., Parati, G., Castiglioni, P., Di Rienzo, M.: Effect of sinoaortic denervation on frequency-domain estimates of baroreflex sensitivity in conscious cats. Am. J. Physiol. Heart Circ. Physiol. **276**, H1987–H1993 (1999)
47. Nollo, G., Porta, A., Faes, L., Del Greco, M., Disertori, M., Ravelli, F.: Causal linear parametric model for baroreflex gain assessment in patients with recent myocardial infarction. Am. J. Physiol. Heart Circ. Physiol. **280**, H1830–H1839 (2001)
48. Barbieri, R., Triedman, J.K., Saul, J.P.: Heart rate control and mechanical cardiopulmonary coupling to assess central volume: a systems analysis. Am. J. Physiol. Regul. Integr. Comp. Physiol. **283**, R1210–R1220 (2002)

49. Faes, L., Porta, A., Cucino, R., Cerutti, S., Antolini, R., Nollo, G.: Causal transfer function analysis to describe closed loop interactions between cardiovascular and cardiorespiratory variability signals. Biol. Cybern. **90**, 390–399 (2004)
50. Faes, L., Pinna, G.D., Porta, A., Maestri, R., Nollo, G.: Surrogate data analysis for assessing the significance of the coherence function. I.E.E.E. Trans. Biomed. Eng. **51**, 1156–1166 (2004)
51. Giddens, D.P., Kitney, R.I.: Neonatal heart-rate variability and its relation to respiration. J. Theor. Biol. **113**, 759–780 (1985)
52. Patzak, A., Lipke, K., Orlow, W., Mrowka, R., Stauss, H., Windt, E., et al.: Development of heart rate power spectra reveals neonatal peculiarities of cardiorespiratory control. Am. J. Physiol. Regul. Integr. Comp. Physiol. **271**, R1025–R1032 (1996)
53. Rehan, V.K., Fajardo, C.A., Haider, A.Z., Alvaro, R.E., Cates, D.B., Kwiatkowski, K., et al.: Influence of sleep state and respiratory pattern on cyclical fluctuations of cerebral blood flow velocity in healthy preterm infants. Biol. Neonate. **69**, 357–367 (1996)
54. Mazursky, J.E., Birkett, C.L., Bedell, K.A., Ben-Haim, S.A., Segar, J.L.: Development of baroreflex influences on heart rate variability in preterm infants. Early Hum. Dev. **53**, 37–52 (1998)
55. Schechtman, V.L., Henslee, J.A., Harper, R.M.: Developmental patterns of heart rate and variability in infants with persistent apnea of infancy. Early Hum. Dev. **50**, 251–262 (1998)
56. Rosenstock, E.G., Cassuto, Y., Zmora, E.: Heart rate variability in the neonate and infant: analytical methods, physiological and clinical observations. Acta Paediatr. **88**, 477–482 (1999)
57. Valimaki, I., Rantonen, T.: Spectral analysis of heart rate and blood pressure variability. Clin. Perinatol. **26**, 967–980 (1999)
58. von Siebenthal, K., Beran, J., Wolf, M., Keel, M., Dietz, V., Kundu, S., et al.: Cyclical fluctuations in blood pressure, heart rate and cerebral blood volume in preterm infants. Brain and Development. **21**, 529–534 (1999)
59. Griffin, M.P., Moorman, J.R.: Toward the early diagnosis of neonatal sepsis and sepsis-like illness using novel heart rate analysis. Pediatrics. **107**(1), 97–104 (2001)
60. Longin, E., Schaible, T., Lenz, T., Konig, S.: Short term heart rate variability in healthy neonates: normative data and physiological observations. Early Hum. Dev. **81**, 663–671 (2005)
61. Longin, E., Gerstner, T., Schaible, T., Lenz, T., Konig, S.: Maturation of the autonomic nervous system: differences in heart rate variability in premature vs. term infants. J. Perinat. Med. **34**, 303–308 (2006)
62. Khattak, A.Z., Padhye, N.S., Williams, A.L., Lasky, R.E., Moya, F.R., Verklan, M.T.: Longitudinal assessment of heart rate variability in very low birth weight infants during their NICU stay. Early Hum. Dev. **83**, 361–366 (2007)
63. Lewicke, A., Corwin, M., Schuckers, M., Xu, X., Neuman, M., Schuckers, S.: Analysis of heart rate variability for predicting cardiorespiratory events in infants. Biomed. Signal Process. Control. **7**(4), 325–332 (2012)
64. Gerstein, G.L., Mandelbrot, B.: Random walk models for the spike activity of a single neuron. J. Biophys. **4**, 41–68 (1964)
65. Daley, D., Vere-Jones, D.: An Introduction to the Theory of Point Processes. Springer, New York (1988)
66. Ogata, Y.: Statistical models for earthquake occurrences and residual analysis for point processes. J. Am. Stat. Assoc. **83**, 9–27 (1988)
67. Chhikara, R.S., Folks, J.L.: The Inverse Gaussian Distribution: Theory, Methodology, and Applications. Marcel Dekker, New York (1989)
68. Poisson Distribution: (https://en.wikipedia.org/wiki/Poisson_distribution)
69. Exponential Distribution: (https://en.wikipedia.org/wiki/Exponential_distribution)
70. Gamma Distribution: (https://en.wikipedia.org/wiki/Gamma_distribution)
71. Log-normal Distribution: (https://en.wikipedia.org/wiki/Log-normal_distribution)
72. Inverse Gaussian Distribution: (https://en.wikipedia.org/wiki/Inverse_Gaussian_distribution)
73. Pawitan, Y.: All Likelihood: Statistical Modelling and Inference Using Likelihood. Oxford, London (2001)

74. Brown, E.N., Barbieri, R., Ventura, V., Kass, R.E., Frank, L.M.: The time-rescaling theorem and its application to neural spike train data analysis. Neural Comput. **14**, 325–346 (2002)
75. Brown, E.N., Barbieri, R., Eden, U.T., Frank, L.M.: Likelihood methods for neural data analysis. In: Feng, J. (ed.) Computational Neuroscience: A Comprehensive Approach, pp. 253–286. CRC, London (2003)
76. Brown, E.N., Barbieri, R., Ventura, V., et al.: The time-rescaling theorem and its application to neural spike train data analysis. Neural Comput. **14**(2), 325–346 (2002)
77. Barbieri, R., Matten, E.C., Alabi, A.A., Brown, E.N.: A point-process model of human heartbeat intervals: new definitions of heart rate and heart rate variability. Am. J. Physiol. Heart Circ. Physiol. **288**(1), H424–H435 (2005)
78. Barbieri, R., Brown, E.N.: Analysis of heartbeat dynamics by point process adaptive filtering. I.E.E.E. Trans. Biomed. Eng. **53**(1), 4–12 (2006)
79. Barbieri, R., Brown, E.N.: Application of dynamic point process models to cardiovascular control. Bio Systems. **93**(1–2), 120–125 (2008)
80. Chen, Z., Brown, E.N., Barbieri, R.: Assessment of autonomic control and respiratory sinus arrhythmia using point process models of human heartbeat dynamics. I.E.E.E. Trans. Biomed. Eng. **56**, 1791–1802 (2009)
81. Chen, Z., Brown, E.N., Barbieri, R.: Characterizing nonlinear heart-beat dynamics within a point process framework. I.E.E.E. Trans. Biomed. Eng. **57**(6), 1335–1347 (2010)
82. Indic, P., Paydarfar, D., Barbieri, R.: A point process model of respiratory dynamics in early physiological development. Proc. IEEE Eng. Med. Biol. Soc. Conf. **2011**, 3804–3807 (2011)
83. Indic, P., Bloch-Salisbury, E., Bednarek, F., Brown, E.N., Paydarfar, D., Barbieri, R.: Assessment of cardio-respiratory interactions in preterm infants by bivariate autoregressive modeling and surrogate data analysis. Early Hum. Dev. **87**(7), 477–487 (2011)
84. Indic, P., Paydarfar, D., Barbieri, R.: Point process modeling of inter-breath interval: a new approach for the assessment of instability of breathing in neonates. I.E.E.E. Trans. Biomed. Eng. **60**(10), 2858–2866 (2013)
85. Gee, A.H., Barbieri, R., Paydarfar, D., Indic, P.: Uncovering statistical features of bradycardia severity in premature infants using a point process model. Conf. Proc. IEEE Eng. Med. Biol. Soc. **2015**, 5855–5858 (2015)
86. Gee, A.H., Barbieri, R., Paydarfar, D., Indic, P.: Improving heart rate estimation in preterm infants with bivariate point process analysis of heart rate and respiration. Conf. Proc. IEEE Eng. Med. Biol. Soc. **2016**, 920–923 (2016)
87. Gee, A.H., Barbieri, R., Paydarfar, D., Indic, P.: Predicting bradycardia in preterm infants using point process analysis of heart rate. I.E.E.E. Trans. Biomed. Eng. **64**(9), 2300–2308 (2017)
88. Pini, N., Lucchini, M., Fifer, W.P., Signorini, M.G., Barbieri, R.: A point process framework for the characterization of sleep states in early infancy. Conf. Proc. I.E.E.E. Eng. Med. Biol. Soc. **8857555**, 3645–3648 (2019)

Chapter 9
Back to the Future: Prenatal Life and Perinatal Programming

Flaminia Bardanzellu and Vassilios Fanos

Contents

In God we trust. All the others please bring data (W.E. Deming)

9.1 Introduction

The concept of big data introduces to the era of data-driven medicine. The medicine of the future will provide infinite possibilities, both depending on the introduction of several innovative technologies, and on the step toward artificial intelligence.

Looking at the new progresses in medical field, we could propose numerous questions requiring deep reflections:

- "From contact lenses to heart valves: are we already cyborgs?"
- "Will bioengineering create a new race of humans?"
- "What if artificial intelligence knew what you wanted before you did?"
- "Would you have an intimate relationship with a robot?"

F. Bardanzellu (✉) · V. Fanos
Neonatal Intensive Care Unit, Department of Surgical Sciences, AOU University of Cagliari, Cagliari, Italy

© Springer Nature Switzerland AG 2021
D. Pani et al. (eds.), *Innovative Technologies and Signal Processing in Perinatal Medicine*, https://doi.org/10.1007/978-3-030-54403-4_9

- "Are humans the reproductive organs of technology?"
- "To what extent should we genetically manipulate living organism?"
- "Should parents be allowed to choose their children?"

These are arguments that we would never thought a few years ago, and are the sign of the evolution of technologies applied to medical research.

In the last century, we assisted at the transition from intuition-based medicine to evidence- and precision-based medicine. In this evolution, the holistic view of human being as a system biology and the introduction of "omics" technologies gave a significant contribution.

Biomedicine reached an inflection point with the introduction of "omics" tools and big data, implying a dramatic increase in our knowledge; in fact, we are moving from a descriptive and reductive approach to an integrated approach that will interpret massive data networks obtained from huge patients' cohorts, healthy patients, and experimental organisms to determine physiologic and pathologic processes specific for each individual [53]. This will mean the introduction of "artificial intuition", decoding the "black box" of medicine.

In the era of "imprecision medicine", where it has been evidenced that in the USA, the second most used drug is effective only in one out of twenty-five patients [82], metabolomics could offer accuracy and precision.

A holistic view cannot be obtained without integrating information from genome, transcriptome, proteome, metabolome, microbiome, epigenome, exposome, etc.

In the last years, "omics" technologies provided an exceptional tool to investigate the composition of many fluids and tissues. These technologies allowed a detailed description of selective gene expression, microbiota characteristics, multipotent stem cells, and dynamic changes in biofluids.

The importance of epigenetics ("epi" = being over; "genetics" = genetics), meaning what can influence genic expression, is highlighted by the observation that identical twins are not identical! Despite an identical genome, the influence played by a different epigenome determines a different phenotype.

Recognized the importance of microbiome, currently investigated through numerous research and scientific publications, the role of less known acetylome, methylome, virome, fungome, cytokinome, chimerismome, etc. will be, in the future, better clarified.

9.2 Perinatal Programming

The concept of perinatal programming is defined as the response by a developing organism to a specific challenge during a critical time window that alters the trajectory of development qualitatively and/or quantitatively with resulting persistent effects on phenotype. In the last years, according to such concept, it has been evidenced that an inadequate intrauterine and early perinatal environment can affect fetal and neonatal life and can also determine long-term negative consequences.

Recent and intriguing titles were published on this topic, such as

- "Individualized medicine from prewomb to tomb" [88]
- "From the cradle to the grave: the early-life origins of chronic obstructive pulmonary disease" [83]
- "David Barker: the revolution that anticipates existence" [46]

According to the concept of pre- and perinatal programming, health and disease would be influenced by factors such as pre- and peri-conceptional environment, preterm birth, intrauterine growth restriction (IUGR), maternal gestational diabetes, hyperoxia during postnatal life in addition to the events associated with child and adult life, including stress and senescence. Many pre- and perinatal events, not fully known up to now, can determine epigenetic changes on DNA, such as methylation, deacetylation, etc. [50].

There are really interesting concepts, since it is well known that most of the diseases are multigenic and multifactorial, and epigenetics is a highly influencing factor.

It has been observed that specific perturbations (i.e., caloric restriction) occurring during the first, second, or third trimester of pregnancy can be associated with different health outcomes later in life, involving well-defined organs or apparatus according to the time of occurrence (i.e., cardiovascular and metabolic disorders in the first trimester, pulmonary and renal diseases in the second trimester, and neuropsychiatric disorders in the last trimester of pregnancy) [79]; thus, this can be seen in a different perspective, underlining that each window of vulnerability, if well faced and sustained with the right interventions, could be transformed in a window of opportunity to improve the development and the whole life.

And therefore, can the fetus be considered the father of the man? Does the placenta (organ representing the surface of exchange between mother and offspring) be considered the center of the chronic Universe? [86]

It is now clear that the development of several organs is influenced by the early moments of life.

According to Santiago Ramon y Cajal, Spanish pathologist, histologist, neuroscientist, and Nobel laureate, the father of modern neuroscience, "The total arborization of a neuron represents the graphic history of conflicts suffered during the developmental life", since these cells do not undergo regeneration.

Moreover, also in animal models, it has been shown that the suckling pigs' duodenum gains 42% of its weight during the first 24 hours of life [3].

The effects of maternal caloric restriction have also been elegantly described in the review "I'm eating for two", which investigates the effects of uteroplacental insufficiency on the reduction in oxygen and energy supply in the fetus, leading to an activation of fetal mitochondria (energetic stations of the cells) to satisfy the cellular need of energy. This determines an increase in cellular oxidative stress and mitochondrial dysfunction and, subsequently, to long-term problems in the functioning of several organs.

Recently, it has been also evidenced that human brains undergo fetal modifications potentially predisposing to neurodegenerative diseases in adulthood [32].

Fetal brain is highly susceptible, especially in the third trimester of pregnancy, when the main cerebral development occurs. In fact, at 35 weeks of gestational age, it weights about 2/3 of the mature brain.

Factors affecting brain development in fetal and early postnatal life are also represented by alcohol, smoke and aluminum exposure (i.e., through soy-milk formulas, vaccines, infusions for parenteral nutrition, etc.) [35].

Perinatal programming also influences hearth development, as reported in several studies, especially investigating IUGR and prematurity as predisposing factors for cardiovascular long-term homeostasis [33, 67] and kidney maturation, influencing the number of nephrons at birth and therefore neonatal susceptibility to renal disease (up to failure) and susceptibility to nephrotoxic drugs in the future life [34, 43].

In this context, we can also affirm the existence and the power of breast milk-associated perinatal programming. In fact, in the first weeks of life, breast milk (BM) is able to change the fate of newborns' metabolism [25].

In pathological conditions, BM could interfere with neonatal development and can increase the risk of impaired function of several organs in the early life. Such concept seems particularly evident in BM of obese mothers, potentially affected also by gestational diabetes mellitus [49].

Thus, great interest concerns the peculiar role and the characteristics of BM in newborn nutrition and development, influencing his whole life.

Several studies have been already performed in order to evaluate the features of the biofluid naturally predisposed to neonatal growth during the first moments of life, sustaining the delicate phase of adaptation to postnatal life.

In our opinion, the "omics" approach is the best tool to evaluate breast milk mediators in a detailed and dynamic way. Metabolomics, in recent years, highly improved the comprehension of its composition and properties. BM is different among mothers and can modulate its composition according to each newborn's requirements. BM is highly beneficial in the vulnerable category of premature babies [9, 15, 84].

9.3 Metabolomics

Metabolomics is an analytical profiling technique for measuring and comparing large numbers of metabolites of low molecular weight present in biological samples. The terms "metabolism" and "metabolomics" share as their root the ancient Greek word, *metabol*, which means change, and obviously both terms are equally applicable to cell, tissue, or whole organism.

Object of metabolomics investigation is the total repertoire of small molecules present in cells, tissues, organs, or biological fluids, such as sugars, lipids, small peptides, vitamins, and amino acids.

If genetics is the description of "what could happen" and proteomics and transcriptomics of "what is happening", metabolomics is a unique description of "what really and effectively happened" in that organism/cell or tissue. Studying metabolomics is how to investigate in a personal dumpster.

Environment (lifestyle, age, drugs, etc.) can represent epigenetics influencer modifying metabolome [30].

The traditional laboratory methodologies offer late markers that are not enough sensitive or specific to diagnose a disease.

"Omics" technologies (genomics, transcriptomics, proteomics, and metabolomics) are able to detect the complexity of biological systems, through the simultaneous, and often noninvasive analysis of a large amount of data (the so-called "direct intelligence of data").

Since the year 2000 and due to these technologies we assisted to an evolution/revolution that completely changed our approach to scientific data.

Technologies used in metabolomics are nuclear magnetic resonance spectrometry (^{1}H-NMR), gas chromatography-mass spectrometry (GC-MS,) and liquid chromatography-mass spectrometry (LC-MS).

More than 27.500 articles dealing with metabolomics can be found on PubMed. It is considered one out of the ten technologies that will change the world and, in 2018, 1 Euro out of 18 were spent in medical research for metabolomics.

The link between metabolomics and clinical research has been highlighted by several authors:

Metabolomics: Bridging the gap between pharmaceutical development and population health [87].

Bridging the gap between clinicians and system biologists: from network biology to translational biomedical research [51]

How a metabolomics study can be performed? It should be accurately programmed, regarding the setting, the patients, and the types of samples. The experimental design should include at least an experimental group and a control group. The kind of samples under evaluation should be correctly stored and undergoes processing (^{1}H-NMR, MS). Then, a multivariate statistical analysis is applied and the results are interpreted (scale-free networks) to individuate significant associations [71].

Patti et al. explain how we can obtain results with targeted or untargeted metabolomics. In the first case, we suppose "a priori" the association of a metabolite variation and we confirm or refuse it. On the contrary, untargeted metabolomics investigates all the metabolites, to state "a posteriori" hypotheses from obtained results.

Samples potentially used in prenatal and perinatal metabolomics analysis can be collected from mothers (amniotic fluid, placenta, blood, urine, breast milk, erythrocytes, hair, and vaginal secretions) or from the neonate (urine, blood, saliva, bronchoalveolar fluid, exhaled air condensate, stools, and umbilical cord) [71].

In the last years, our research group applied metabolomics in a large number of studies, including the first study available in the literature evaluating BM metabolome and the first review on this topic. Moreover, we investigated urinary metabolic profiles in several obstetrical conditions (maternal obesity, gestational diabetes mellitus, intrauterine growth restriction, chorioamnionitis, congenital CMV infection), perinatal conditions (vaginal delivery, prematurity, nutrition regimen, perinatal asphyxia, sepsis, necrotizing enterocolitis, bronchopulmonary dysplasia, patent ductus arteriosus), and pediatric diseases (autism and nephrourinary disorders).

Below, we report the main results of our experience.

9.3.1 Prenatal Conditions

Metabolomics seems promising in the detection of intrauterine disorders. A restriction in intrauterine growth (IUGR) can highly impair fetal outcome.

Our research group applied metabolomics to investigate and compare urinary samples at birth among IUGR, appropriate for gestational age (AGA) and large for gestational age (LGA) neonates (often associated with diabetic mothers). Different metabolites characterized the urines of these categories. In detail, an increase in myoinositol was reported in IUGR neonates [7, 23], and was also confirmed at four days of life [22], and at one week [16]. Variable products are correlated to energy production, antioxidant activity, and brain development.

Metabolic profiles of IUGR and LGA resulted similar at birth, although they represent opposite phenotypes: a reduced and an excessive fetal nutrition, influenced by maternal nutrition, seem associated with the same metabolic footprints, potentially leading to similar metabolic long-lasting effects [16, 23].

Maternal obesity seems to influence neonatal metabolism too. We evaluated this effect through the analysis of maternal placental samples on normal-weight mothers versus obese mothers. As a result, several metabolites resulted different, also taking into account maternal gestational diabetes mellitus. These metabolites were mostly involved in antioxidant activity and energy production. In obese mothers, there was a reduction in lipids for fetal development, impairing its metabolism [49].

Intra-uterine infections can affect neonatal development and outcome, being also associated with premature delivery. Our group evaluated urinary samples from neonates born from mothers affected by chorioamnionitis, demonstrating that the inflammatory state influenced metabolic pathways, especially in relation to mitochondrial dysfunction [48].

Maternal infection by human cytomegalovirus (HCMV) can be congenitally acquired by the fetus during intrauterine life, and potentially leads to fetal damage or even death [39]. With metabolomics, we were able to separate urinary samples, at birth, of those neonates showing clinical signs of HCMV after fetal infection. In detail, symptomatic patients showed an increase in metabolites representing intermediates of compensative mechanisms to face infection-related energy deficiency, metabolites involved in cell volume regulation and in viral metabolism [39].

Moreover, HCMV infection also seems to influence amniotic fluid (AF) metabolic composition, as highlighted by a more recent study. In fact, metabolomics allowed the separation of AF samples from mothers who transmitted HCMV to their fetuses than those who have not transmitted it and healthy controls, also allowing the individuation of those congenitally infected newborns showing clinical signs at birth [47].

These findings, although preliminary, evidence that metabolomics is a reliable and promising tool to individuate those neonates congenitally infected or more susceptible to develop clinical manifestations, improving a precocious diagnosis of such infection [8, 11, 47, 59].

Metabolomics has been also applied in obstetrical conditions, helping in the detection of the relations between premature rupture of membranes (PROM) and labor onset, through maternal urinary samples [66].

9.3.2 Perinatal Conditions

Interesting findings have been also obtained applying metabolomics in the analysis of the delivery mode. This is useful to investigate the peculiar physiology associated with birth modality and potentially influencing extra-uterine life adaptation and neonatal outcome.

Urinary profiles of neonates born by cesarean section (CS) were different than vaginal delivery (VD), especially regarding metabolites involved in thermoregulation at birth and energy metabolism, and metabolites of bacterial origin (due to a different bacterial colonization in the two different delivery modalities) [65].

In the field of premature birth, condition that highly impairs neonatal outcome especially according to prematurity degree, our research group performed several studies.

Urinary samples collected at birth have been compared among full-term and preterm babies, detecting different metabolic profiles potentially associated with preterm birth and impairing the future outcome.

As a result, amino acid's metabolism seems the most involved in postnatal development and is different according to maturity at birth [5].

According to a more recent study, prematurity determined an increase in fat mass percentage and most varying metabolites were those associated with energy production and antioxidant activity [68].

Among the causes of neonatal death or severe disability (up to cerebral palsy), perinatal asphyxia is common [60]. According to our results, metabolomics seems promising to distinguish urinary samples in relation to the asphyxia-related outcome. This has been shown on a small sample of three patients, highly different according to cerebral damage and outcome [41]. Successively, on larger groups, metabolomic urinary differences were reported after 48 hours of life (especially metabolites involved in energy demand, kidney damage, and deficient oxidative metabolism) [61], and at the end of hypothermic treatment (72 hours), at a week and a month after birth, well describing the damage progression [60, 76].

Neonatal sepsis can be caused by viral, fungal, or bacterial infections. It is particularly dangerous, especially in premature neonates. "Early-onset" sepsis (EOS) occurs within 72 hours of life, while "late-onset" sepsis (LOS) between 72 hours and 6 days of life. An early diagnosis can improve management, survival, and outcome; even if, currently, precocious and sensible biomarkers are still lacking [62, 73, 81, 85] and metabolomics could improve the diagnosis and the monitoring of antibiotic therapy [24, 70, 74].

We reported significant differences between urinary samples of neonates affected by sepsis (EOS and LOS) and healthy controls [40], even in fungal sepsis, where D-serine level variation also helped in predicting therapy response [24]. The increase in some amino acids could be related to hypermetabolic and hypercatabolic state during sepsis, determined by the increase in energy demand.

Studies on sepsis have been also performed on pediatric patients. The objectives of metabolomics in sepsis are the specific individuation of etiology, the prediction of the severity, the potential individuation of therapy responders, and monitoring of drug toxicity.

9.3.3   Post-Natal Period

Among the diseases affecting neonates and potentially impairing their outcome, we focused our studies on necrotizing enterocolitis (NEC), bronchopulmonary dysplasia (BPD), and patent ductus arteriosus (PDA).

NEC is a severe bowel inflammation that can occur in premature neonates. Its pathogenesis is not fully clarified and symptoms can be not specific. The detection of precocious and reliable biomarkers could be very useful [26, 77].

Thus, through the application of metabolomics, we described specific urinary pathways few days before and at the moment of the disease clinical onset (instead of healthy controls), suggesting the existence of precocious biomarkers [21, 77]. In detail, since gluconic acid and variations in carbohydrates' metabolism were the characterizing metabolite increased in patients affected by NEC, metabolomics seems useful in the comprehension of NEC and in the early detection of the onset and progression of the disease.

BPD, which is the supplemental oxygen requirement at 36 weeks of post-menstrual age, represents a severe lung disease impairing pulmonary development of premature neonates [6, 57], frequently leading to death of very low birth weight neonates [78].

In this field, we speculated the characterization of different metabolomic pathways in the urines, at birth, of those premature newborns subsequently developing BDP. Among three studies, our research group demonstrated a clear separation between BPD patients and healthy controls, both at birth and in the first week of life, potentially anticipating the individuation of BDP susceptibility and underlying that BDP pathogenesis could start in the early postnatal life [42, 63, 78].

It could be very useful to individuate specific biomarkers correlating with clinical features and prognosis of mild, moderate, and severe BPD potentially predicting the outcome and therapy response, with the aim of an individualized treatment.

9.3.4   *Breastfeeding*

The first metabolomic study investigating BM composition was performed by our group in 2012. We evaluated samples collected from mothers delivering preterm (26–36 weeks of gestational age) and full-term neonates during lactation and compared them with commercial formula milk (FM).

BM showed more lactose, while maltose was higher in FM. Some differences also characterized fatty acids and a progressive metabolic change characterized BM maturation, especially regarding carbohydrate profile and lactose, that increased during lactation (milk maturation) [15].

Among BM components, a pivotal role is played by BM's oligosaccharides (HMO), resulting beneficial for several functions. They can modulate gut neonatal microbiome (increasing commensal species), immune response and can reduce the susceptibility to numerous pathologies [10, 44]. BM oligosaccharides (levels and quality) are influenced by several factors. Among these, the most relevant is maternal genetics; in detail, the expression of two specific genes (*Se* and *Le* genes) determines the secretory phenotype (*Se+*) or non-secretory (*Se-*). Consequently, BM from *Se* + mothers contains some fucosylated oligosaccharides that can protect the newborn from infections and necrotizing enterocolitis (NEC), and therefore improve neonatal health [44]. Our group studied BM metabolomic profile and demonstrated a clear separation between BM samples from *Se* + mothers and *Se-* mothers. Such an analysis, if currently performed, could help in the individuation of newborns more susceptible to infections (according to maternal genetic factors), since they are less protected by maternal HMOs. This could potentially improve BM supplementation strategies [27].

We also published interesting results on neonatal urinary metabolome according to the neonatal nutrition regimen. Neonatal urines were evaluated during the first week of life and the diet with BM or FM was reported. In detail, at birth urinary metabolome was comparable in neonate born too small and too large, characterized by the increase of mediator myoinositol and by the laboratory finding of serum hypoglycemia. Subsequently, it has been shown that, after a week of life, the urinary metabolome was influenced by the milk assumed, independently by the weight showed at birth [25]. It can be deduced that BM in the first week of life can totally change neonatal metabolism and modulate organ development. Similar findings were also reported at 130 days of postnatal life, because of the changes in urinary metabolome according to the dietary regimen [17].

Metabolic urinary variations mostly regarded energy production, antioxidant, brain and pulmonary development, and gut microbe-derived metabolites.

In a recent study, our group detected significant differences among BM samples from mothers delivering preterm twins or triplets instead of preterm singletons; the higher protein content (and the reduced lactose) found in BM from mothers delivering preterm multiples could face the needs of such vulnerable category, promoting and sustaining growth and organ development [20].

Metabolomics can also individuate drugs in maternal milk [38].

Finally, breast milk has been identified as an incredible source of maternal stem cells, able to be transmitted to the neonatal gut via breastfeeding and able to migrate in neonatal organs, especially the brain, becoming cerebral cells [52, 55].

These studies underline the importance of breastfeeding short- and long-term health, and therefore the importance of a targeted diet in the early stages of life.

Perinatal programming is a field requiring deep investigations, since pediatric health starts in intrauterine and perinatal life.

9.3.5   Children

The group of autistic spectrum disorders (ASD) includes various psychiatric illnesses determined by a genetic predisposition in addition to numerous epigenetic components.

Despite a great increase in the diagnosis of such disorders in recent decades, unique biological markers are not currently available, and the diagnosis is still largely based on the patient's symptoms.

According to recent metabolomics studies, relevant epigenetic factors would be constituted by oxidative mechanisms as well as by some peculiar modifications of the intestinal microbiome of affected patients, which would involve changes in the gut-brain interaction axis. In this sector, our group, along with other researchers, has highlighted peculiar profiles in the urinary metabolomics of autistic patients, paving the way to what could be an innovative approach to this psychiatric disorder. Moreover, it could provide a metabolic footprint characterizing the aforementioned condition, helping to understand its pathogenesis, potentially favoring early and sensitive diagnoses, a possible monitoring of disease progression and suggesting new potential therapeutic targets.

In detail, autistic children showed a urinary increase in metabolites involved in oxidative mechanisms, products of carbohydrate metabolism, bacterial metabolites suggesting an increase in *Clostridia* spp. and changes in the levels of precursors of neurotransmitters. As a result, the importance of the diet in these children is underlined [64, 75].

Moreover, we characterized the metabolite patterns associated with nephrouropathies, through urine samples obtained from children affected by nephrouropathies compared to healthy controls. As a result, the urine metabolite profiles seem a promising, non-invasive tool in this field, correlating to nephrourological disorders [4].

9.3.6   Adults

Our group also performed a metabolomics study in adult patients (mean age 24 years), comparing the samples of those who were born at term and those born showing extremely low birth weight. The differences detected in the two groups were mostly related to the alterations in the arginine and proline metabolism, in the purine and pyrimidine metabolism in the hystidine, beta-alanine metabolism, and in the urea cycle [36].

Promising results have been obtained also in oncology and in other fields of adults' medicine [72].

9.4   Microbiomics

Metabolomics is also defined as the "Rosetta Stone" of Microbiomics, meaning that metabolites detected are often produced by bacteria and can help in decoding the interactions occurring by bacterial species and the host.

The concept that "We live in the Age of Bacteria (as it was in the beginning, is now, and ever shall be, until the world ends" (Stephen Jay Gould, Cambridge, MA, 1993) is known since long time.

The Nobel Prize in Medicine 1958 was assigned to Joshua Lederberg "for his discoveries concerning genetic recombination and the organization of the genetic material of bacteria (microbiome)".

Moreover, Jaroes Raes (The Flanders Institute of Biology) introduced the concept of HOMO BACTERIENS, since in his opinion "You are not human, you are a walking bacterial colony".

Currently, a quick search for "Microbiome" in scientific journals online demonstrates how significantly this field of research has been growing over the past 10 years.

Giving some numbers, gut microbiota can weight about 1–2 Kg, 95% of bacteria are located in the gut and intestinal surface is about 400 m².

Heinz et al. introduced the concept of superorganism (Man + Microbes) and superorgan (Gut + Microbiota), allowing the thought that we are What we Host! [54].

The current opinion describes the brain and the gut as the first and the second brain of our body. These structures seem communicate through biological signals (mostly short chain fatty acids – SCFAs).

Intestinal microbiota is mostly represented by 3 phyla: Bacteroidetes, Actinobacteria, and Firmicutes [69] in different percentages; thus, our microbiota is a unique fingerprint.

Some species are associated with several diseases, such as *Clostridium* spp. that seems involved in preeclampsia, NEC, inflammatory bowel diseases, autism, and stroke.

Modifications in the microbiome have been also associated with the onset of sepsis [2].

This should make us reflect on the consequences of uncontrolled antibiotic use [13].

Germ-free rats do not gain fat and assume the microbiome of the donor [19]. Microbiota modifications during life are due to pre- and perinatal factors, lifestyle, habits, nutrition, antibiotics, and many others environmental factors (in addition to development itself and hormonal changes).

The womb is sterile? In the last years appeared the evidence of a placental microbiome [1], which seems similar to maternal oral microbiome (periodontitis). In fact, such inflammatory situation could be linked to preterm birth.

Even mode of delivery can influence neonatal microbiome [14, 28, 80]. Neonates born by cesarean section seem associated with a higher level of *Clostridia* spp., less bifidobacteria, bacteroides, and lactobacilli [80].

According to Dominguez-Bello, neonates born by CS show maternal skin microbiome, while those born by vaginal delivery are colonized by the maternal vaginal microbiome.

A lactating neonate assumes about 800 ml of BM daily, ingesting 100.000–10.000.000 bacteria (Lactobiome).

This influences neonatal gut microbiome ([29], Le Doare et al. [58], Bazanella et al. [12], [18], Fanos [38], Reali 2018), potentially influencing neonatal health [31].

In the future, several beneficial applications will be obtained by fecal microbiome transplantation. This consists in the infusion of feces from a healthy donor into the gut of a recipient to cure a specific disease (by nasogastric or nasoduodenal tube, colonoscope, enema, or capsule). The high success rate and safety in the short term is reported for recurrent *C. difficile* infection. It could be beneficial for a wide range of disorders, including Parkinson's disease, fibromyalgia, chronic fatigue syndrome, myoclonus dystopia, multiple sclerosis, obesity, insulin resistance, metabolic syndrome, and autism. Even if there are many unanswered questions [56].

9.5 Conclusions

Genome is highly influenced by several factors. Among them, we underline the role of nutrition and drugs, open the way to nutrimetabolomics, nutrimicrobiomics, pharmametabolomics, etc.

We believe that the translational power of metabolomics in the next future will be represented by the whole comprehension of the meaning of each metabolite significant variation, understanding the cause of variation, the effects played on fetus and newborn, and if metabolites could become therapeutic targets. Moreover, correlating metabolites' levels with neonatal specific pathways and health outcome will help to adequate supplementation of nutrition strategies in the most vulnerable category, supplying to each neonate specific deficiency, in the perspective of a personalized nutrition.

In conclusion, all the organisms show a great interindividual variability (basal and after stimulus) and each one is characterized by an intrinsic fragility and resilience.

It will be useful to individuate the mostly varying metabolites in neonatology [45]. Thanks to such specific and precise information, innovative tools could be introduced, such as simple urinary sticks coupled to metabolomics detection to perform rapid diagnosis.

Metabolomics and microbiomics could bring to the medicine of the future [37] and to the 10P pediatrics:

- *Personalized*
- *Prospective*
- *Predictive*
- *Preventive*
- *Precise*
- *Participated*
- *Patient-centered*
- *Psycocognitive*
- *Postgenomic*
- *Public*

References

1. Aagaard, K., Ma, J., Antony, K.M., et al.: The placenta harbors a unique microbiome. Sci. Transl. Med. **6**, 237ra65 (2014)
2. Alverdy, J.C., Krezalek, M.A.: Collapse of the microbiome, emergence of the Pathobiome, and the immunopathology of Sepsis. Crit. Care Med. **45**, 337–347 (2017)
3. Ashwell, M.: Widdowson CH 1906–2000. Biogr. Mems. R. Soc. London. **48**, 483–506 (2002)
4. Atzori, L., Antonucci, R., Barberini, L., et al.: 1H NMR-based metabolic profiling of urine from children with nephrouropathies. Front. Biosci. (2), 725–732 (2010)
5. Atzori, L., Antonucci, R., Barberini, L., et al.: 1H NMR-based metabolomic analysis of urine from preterm and term neonates. Front. Biosci. **3**, 1005–1012 (2011)
6. Baraldi, E., Giordano, G., Stoccchero, M., et al.: Untargeted metabolomic analysis of amniotic fluid in the prediction of preterm delivery and bronchopulmonary dysplasia. PLoS One. **11**, e0164211 (2016)
7. Barberini, L., Noto, A., Fattuoni, C., et al.: Urinary metabolomics (GC-MS) reveals that low and high birth weight infants share elevate dinositol concentrations at birth. J. Matern. Fetal Neonatal Med. **27**, 20–26 (2014)
8. Barberini, L., Noto, A., Saba, L., et al.: Multivariate data validation investigating primary HCMV infection in pregnancy. Data Brief. **9**, 220–230 (2016)
9. Bardanzellu, F., Fanos, V., Reali, A.: "Omics" in human colostrum and mature milk: looking to old data with new eyes. Nutrients. **9**, 843 (2017)
10. Bardanzellu, F., Fanos, V., Strigini, F.A.L., et al.: Human breast milk: exploring the linking ring among emerging components. Front. Pediatr. **6**, 215 (2018)
11. Bardanzellu, F., Fanos, V., Reali, A.: Human breast milk-acquired Cytomegalovirus infection: certainties, doubts and perspectives. Curr. Pediatr. Rev. **15**, 30–41 (2019)
12. Bazanella, M., Maier, T.V., Clavel, T., et al.: Randomized controlled trial on the impact of early-life intervention with bifidobacteria on the healthy infant fecal microbiota and metabolome. Am. J. Clin. Nutr. **106**, 1274–1286 (2017)
13. Blaser, M.J.: Missing Microbes: How the Overuse of Antibiotics Is Fueling Our Modern Plagues, First edn. Editor Henry Holt and Co. New York City (2014)
14. Bokulich, N.A., Chung, J., Battaglia, T., et al.: Antibiotics, birth mode, and diet shape microbiome maturation during early life. Sci. Transl. Med. **8**, 343ra82 (2016)

15. Cesare Marincola, F., Noto, A., Caboni, P., et al.: A metabolomic study of preterm human and formula milk by high resolution NMR and GC/MS analysis: preliminary results. J. Matern. Fetal Neonatal Med. **25**, 62–67 (2012)
16. Cesare Marincola, F., Dessì, A., Pattumelli, M.G., et al.: (1)H NMR-based urine metabolic profile of IUGR, LGA, and AGA newborns in the first week of life. Clin. Chim. Acta. **451**, 28–34 (2015)
17. Cesare Marincola, F., Corbu, S., Lussu, M., et al.: Impact of early postnatal nutrition on the NMR urinary metabolic profile of infant. J. Proteome Res. **15**, 3712–3723 (2016)
18. Charbonneau, M.R., Blanton, L.V., Di Giulio, D.B., et al.: A microbial perspective of human developmental biology. Nature. **535**, 48–55 (2016)
19. Clarke, G., Stilling, R.M., Kennedy, P.J., et al.: Minireview: gut microbiota: the neglected endocrine organ. Mol. Endocrinol. **28**, 1221–1238 (2014)
20. Congiu, M., Reali, A., Deidda, F., et al.: Breast milk for preterm multiples: more proteins, less lactose. Twin Res. Hum. Genet. **22**(4), 265–271 (2019)
21. De Magistris, A., Corbu, S., Cesare Marincola, F., et al.: NMR-based metabolomics analysis of urinary changes in neonatal necrotizing enterocolitis. Selected abstracts of the 11th International Workshop on Neonatology, Cagliari (Italy), October 26–31, 2015. ABS 34. J. Pediatr. Neonat. Individual Med. **4**, e040250 (2015)
22. Dessì, A., Atzori, L., Noto, A., et al.: Metabolomics in newborns with intrauterine growth retardation (IUGR): urine reveals markers of metabolic syndrome. J. Matern. Fetal Neonatal Med. **24**, 35–39 (2011)
23. Dessì, A., Marincola, F.C., Pattumelli, M.G., et al.: Investigation of the H-NMR based urine metabolomic profiles of IUGR, LGA and AGA newborns on the first day of life. J. Matern. Fetal Neonatal Med. **27**, 13–19 (2014)
24. Dessì, A., Liori, B., Caboni, P., et al.: Monitoring neonatal fungal infection with metabolomics. J. Matern. Fetal Neonatal Med. **2**, 34–38 (2014)
25. Dessì, A., Murgia, A., Agostino, R., et al.: Exploring the role of different neonatal nutrition regimens during the first week of life by urinary GC-MS metabolomics. Int. J. Mol. Sci. **17**, 265 (2016)
26. Dessì, A., Pintus, R., Marras, S., et al.: Metabolomics in necrotizing enterocolitis: the state of the art. Exp. Rev. Mol. Diagn. **16**, 1053–1058 (2016)
27. Dessì, A., Briana, D., Corbu, S., et al.: Metabolomics of breast Milk: the importance of phenotypes. Meta. **8pii**, E79 (2018)
28. Dominguez-Bello, M.G., Costello, E.K., Contreras, M., et al.: Delivery mode shapes the acquisition and structure of the initial microbiota across multiple body habitats in newborns. Proc. Natl. Acad. Sci. U. S. A. **107**, 11971–11975 (2010)
29. Donovan, S.M., Wang, M., Li, M., et al.: Host-microbe interactions in the neonatal intestine: role of human milk oligosaccharides. Adv. Nutr. **3**, 450S–455S (2012)
30. Dunn, W.B., Broadhurst, D.I., Atherton, H.J., et al.: Systems level studies of mammalian metabolomes: the roles of mass spectrometry and nuclear magnetic resonance spectroscopy. Chem. Soc. Rev. **40**, 387–426 (2011)
31. Faa, G., Gerosa, C., Fanni, D., et al.: Factors influencing the development of a personal tailored microbiota in the neonate, with particular emphasis on antibiotic therapy. J. Matern. Fetal Neonatal Med. **2**, 35–43 (2013)
32. Faa, G., Marcialis, M.A., Ravarino, A., et al.: Fetal programming of the human brain: is there a link with insurgence of neurodegenerative disorders in adulthood? Curr. Med. Chem. **21**, 3854–3876 (2014)
33. Faa, A., Ambu, R., Faa, G., et al.: Perinatal heart programming: long-term consequences. Curr. Med. Chem. **21**, 3165–3172 (2014)
34. Fanni, D., Gerosa, C., Nemolato, S., et al.: "physiological" renal regenerating medicine in VLBW preterm infants: could a dream come true? J. Matern. Fetal Neonatal Med. **25**, 41–48 (2012)

35. Fanni, D., Ambu, R., Gerosa, F., et al.: Aluminium exposure and toxicity in neonates: a practical guide to halt aluminium overload in the prenatal and perinatal periods. Worlds J. Pediatr. **10**, 101–107 (2014)
36. Fanos V, Barberini L, Antonucci R, et al. Metabolomics in neonatology and pediatrics. Clin. Biochem. 2011;44:452–4
37. Fanos, V.: Pediatric and neonatal individualized medicine: care and cure for each and everyone. J. Pediatr. Neonatal Individual. Med. **1**, 7–10 (2012)
38. Fanos, V., Barberini, L., Antonucci, R., et al.: Pharma-metabolomics in neonatology: is it a dream or a fact? Curr. Pharm. Des. **18**, 2996–3006 (2012)
39. Fanos, V., Locci, E., Noto, A., et al.: Urinary metabolomics in newborns infected by human cytomegalovirus a preliminary investigation. Early Hum. Dev. **89**, S58–S61 (2013)
40. Fanos, V., Caboni, P., Corsello, G., et al.: Urinary (1)H-NMR and GC-MS metabolomics predicts early and late onset neonatal sepsis. Early Hum. Dev. **90**, S78–S83 (2014)
41. Fanos, V., Noto, A., Caboni, P., et al.: Urine metabolomic profiling in neonatal nephrology. Clin. Biochem. **47**, 708–710 (2014)
42. Fanos, V., Pintus, M.C., Lussu, M., et al.: Urinary metabolomics of bronchopulmonary dysplasia (BPD): preliminary data at birth suggest it is a congenital disease. J. Matern. Fetal Neonatal Med. **27**, 39–45 (2014)
43. Fanos, V., Gerosa, C., Fanni, C., et al.: The kidney of late preterm infants. Ital. J. Pediatr. **40**, A14 (2014)
44. Fanos, V., Reali, A., Marcialis, M.A., et al.: What you have to know about human milk oligosaccharides. J. Pediatr. Neonat. Individual. Med. **7**, e070137 (2018)
45. Fanos, V., Pintus, R., Dessì, A.: Clinical metabolomics in neonatology: from metabolites to diseases. Neonatology. **113**, 406–413 (2018)
46. Farnetani, I., Fanos, V.: David Barker: the revolution that anticipates existence. J. Pediatr. Neonat. Individual. Med. **3**, e030111 (2014)
47. Fattuoni, C., Palmas, F., Noto, A., et al.: Primary HCMV infection in pregnancy from classic data towards metabolomics: an exploratory analysis. Clin. Chim. Acta. **460**, 23–32 (2016)
48. Fattuoni, C., Pietrasanta, C., Pugni, L., et al.: Urinary metabolomic analysis to identify preterm neonates exposed to histological chorioamnionitis: a pilot study. PLoS One. **12**, e0189120 (2017)
49. Fattuoni, C., Mandò, C., Palmas, F., et al.: Preliminary metabolomics analysis of placenta in maternal obesity. Placenta. **61**, 89–95 (2018)
50. Gascoin-Lachambre, G., Buffat, C., Rebourcet, R., et al.: Cullins in human intra-uterine growth restriction: expressional and epigenetic alterations. Placenta. **31**, 151–157 (2010)
51. Jinawath, N., Bunbanjerdsuk, S., Chayanupatkul, M., et al.: Bridging the gap between clinicians and system biologists: from network biology to translational biomedical research. J Transl. Med. **14**, 324 (2016)
52. Hassiotou, F., Heath, B., Ocal, O., et al.: Breastmilk stem cell transfer from mother to neonatal organs. FASEB J. **28**, (2014)
53. Hawgood, S., Hook-Barnard, I.G., O'Brien, T.C., et al.: Precision medicine: beyond the inflection point. Sci. Transl. Med. **7**(300), (2015)
54. Heinz, C., Mair, W.: You are what you host: microbiome modulation of the aging process. Cell. **156**, 408–411 (2014)
55. Hosseini, S.M., Talaei-khozani, T., Sani, M., et al.: Differentiation of human breast-milk stem cells to neural stem cells and neurons. Neurol. Res. Int. ID807896 (2014)
56. Hyun, H.C., Young-Seok, C.: Fecal microbiota transplantation: current applications, effectiveness, and future perspectives. Clin. Endosc. **49**, 257–265 (2016)
57. Lal, C.V., Bhandari, V., Ambalavanan, N.: Genomics, microbiomics, proteomics, and metabolomics in bronchopulmonary dysplasia. Semin. Perinatol. **42**, 425–431 (2018)
58. Le Doare, K., Holder, B., Bassett, A., et al.: Mother's milk: a purposeful contribution to the development of the infant microbiota and immunity. Review. Front. Immunol. **9**, 361 (2018)

59. Locci, E., Noto, A., Lanari, M., et al.: Metabolomics: a new tool for the investigation of metabolic changes induced by cytomegalovirus. J. Matern. Fetal Neonatal Med. **26**, 17–19 (2013)
60. Locci, E., Noto, A., Puddu, M., et al.: A longitudinal 1H-NMR metabolomics analysis of urine from newborns with hypoxic-ischemic encephalopathy undergoing hypothermia therapy. Clinical and medical legal insights. PLoS One. **13**(4), e0194267 (2018)
61. Longini, M., Giglio, S., Perrone, S., et al.: Proton nuclear magnetic resonance spectroscopy of urine samples in preterm asphyctic newborn: a metabolomic approach. Clin. Chim. Acta. **444**, 250–256 (2015)
62. Ludwig, K.R., Hummon, A.B.: Mass spectrometry for the discovery of biomarkers of sepsis. Mol. BioSyst. **13**, 648–664 (2017)
63. Lussu, M., Pintus, M.C., Palmas, G.: Metabolomics in bronchopulmonary dysplasia: preliminary results. Early Hum. Dev. **89**, 87 (2013)
64. Lussu, M., Noto, A., Masili, A., et al.: The urinary H-NMR metabolomics profile of an Italian autistic children population and their unaffected siblings. Autism Res. **10**, 1058–1066 (2017)
65. Martin, F.-P., Rezzi, S., Lussu, M., et al.: Urinary metabolomics in term newborns delivered spontaneously or with cesarean section: preliminary data. J. Pediatr. Neonat. Individual. Med. **7**, e070219 (2018)
66. Meloni, A., Palmas, F., Barberini, L., et al.: PROM and labour effects on urinary metabolome: a pilot study. Dis. Markers. **2018**, 1042479 (2018)
67. Mercuro, G., Bassareo, P.P., Flore, G., et al.: Prematurity and low birth weight as new conditions predisposing to an increased cardiovascular risk. Eur. J. Prev. Cardiol. **20**(2), 357–367 (2012)
68. Morniroli, D.A., Dessì, A., Gianni, M.L., et al.: Is the body composition development in premature infants associated with a distinctive nuclear magnetic resonance metabolomic profiling of urine? J. Matern. Fetal Neonatal Med. **15**, 1–9 (2018)
69. Mehta, H., Goulet, P.O., Mashiko, S., et al.: Early-life antibiotic exposure causes intestinal Dysbiosis and exacerbates skin and lung pathology in experimental systemic sclerosis. J. Invest. Dermatol. **137**, 2316–2325 (2017)
70. Mussap, M.: Laboratory medicine in neonatal sepsis and inflammation. J. Matern. Fetal Neonatal Med. **4**, 32–34 (2012)
71. Mussap, M., Antonucci, R., Noto, A., et al.: The role of metabolomics in neonatal and pediatric laboratory medicine. Clin. Chim. Acta. **426**, 127–138 (2013)
72. Mussap, M., Barberini, L., Deidda, S., et al.: A gas chromatography-mass spectrometry (GC-MS) metabolomic approach in human colorectal cancer (CRC): preliminary data on the role of monosaccharides and amino acids. Adv. Transl. Med. **7**(23), 727 (2019)
73. Ng, S., Strunk, T., Jiang, P., et al.: Precision medicine for neonatal sepsis. Front. Mol. Biosci. **5**, 70 (2018)
74. Noto, A., Mussap, M., Fanos, V.: Is 1H NMR metabolomics becoming the promising early biomarker for neonatal sepsis and for monitoring the antibiotic toxicity? J. Chemioter. **26**, 130–132 (2014)
75. Noto, A., Fanos, V., Barberini, L., et al.: The urinary metabolomics profile of an Italian autistic children population and their unaffected siblings. J. Matern. Fetal Neonatal Med. **27**, 46–52 (2014)
76. Noto, A., Pomero, G., Mussap, M., et al.: Urinary gas chromatography mass spectrometry metabolomics in asphyxiated newborns undergoing hypothermia: from the birth to the first month of life. Ann. Transl. Med. **4**, 417 (2016)
77. Palmas, F., De Magistris, A., Noto, A., et al.: Metabolomics study on NEC occurrence by GC-MS. Selected abstracts of the 12th International Workshop on Neonatology, Cagliari (Italy), October 19–22, 2016. ABS 80. J. Pediatr. Neonatal Individual. Med. **5**, e050247 (2016)
78. Pintus, M.C., Lussu, M., Dessì, A., et al.: Urinary 1H-NMR metabolomics in the first week of life can anticipate BPD diagnosis. Oxidative Med. Cell. Longev. **2018**, 7620671 (2018)
79. Roseboom, T.J., Painter, R.C., van Abeelen, A.F., et al.: Hungry in the womb: what are the consequences? Lessons from the Dutch famine. Maturitas. **70**, 141–145 (2011)

80. Rutayisire, E., Huang, K., Liu, Y., et al.: The mode of delivery affects the diversity and colonization pattern of the gut microbiota during the first year of infants' life: a systematic review. BMC Gastroenterol. **16**, 86 (2016)
81. Sarafidis, K., Chatziioannou, A.C., Thomaidou, A., et al.: Urine metabolomics in neonates with late-onset sepsis in a case-control study. Sci. Rep. **7**, 45506 (2017)
82. Schork, N.J.: Personalized medicine: time for one-person trials. Nature. **520**, 609–611 (2015)
83. Sly, P.D., Bush, A.: From the cradle to the grave: the early-life origins of chronic obstructive pulmonary disease. ATS J. **193**(1), (2016)
84. Spevacek, A.R., Smilowitz, J.T., Chin, E.L., et al.: Infant Maturity at Birth Reveals Minor Differences in the Maternal Milk Metabolome in the First Month of Lactation. J. Nutr. **145**, 1698–1708 (2015)
85. Stewart, J.C., Embleton, N.D., Marrs, E.C.L., et al.: Longitudinal development of the gut microbiome and metabolome in preterm neonates with late onset sepsis and healthy controls. Microbiome. **5**, 75 (2017)
86. Thomburg, K.L., Marshall, N.: The placenta is the center of the chronic disease universe. Am. J. Obstet. Gynecol. **213**, 14–20 (2015)
87. Tolstikov, V.: Metabolomics: bridging the gap between pharmaceutical development and population health. Meta. **6**, E20 (2016)
88. Topol, E.J.: Individualized medicine from prewomb to tomb. Cell. **157**, 241–253 (2014)

Index

© Springer Nature Switzerland AG 2021   227
D. Pani et al. (eds.), *Innovative Technologies and Signal Processing in Perinatal
Medicine*, https://doi.org/10.1007/978-3-030-54403-4

Printed in the United States
by Baker & Taylor Publisher Services